Recent Trends in INVERTEBRATES

Vol. 7

H.S.Bhamrah
Kavita Juneja

ANMOL PUBLICATIONS

ANMOL PUBLICATIONS
4378/4B, Ansari Road,
Darya Ganj, New Delhi - 110002.

Recent Trends in Invertebrates

ISBN 81 7041 473 3 (Set)
ISBN 81 7041 474 1 (Vol.I)
ISBN 81 7041 475 X (Vol.II)
ISBN 81 7041 476 8 (Vol. III)
ISBN 81 7041 477 6 (Vol. IV)
ISBN 81 7041 478 4 (Vol. V)
ISBN 81 7041 479 2 (Vol. VI)
ISBN 81 7041 480 6 (Vol. VII)
ISBN 81 7041 481 4 (V0l. VIII)

PRINTED IN INDIA

Published by J.L. Kumar for Anmol Publications and Laser Typeset by D.V. Computers, Delhi, Printed at Efficient Offset Printers, New Delhi.

PREFACE

The present title has been designed to meet the requirements of the undergraduate students. In thc Recent Trends in Invertebrate all phyla have been discussed. The representative types of each phylum have been described in detail. This is followed by the classification of the phylum with a description of its common animals. Recent findings and work have been incorporated in the text. Special care has been taken to make the language as simple and lucid as possible. Simple and accurate diagrams accompany the text. In the interest of the students efforts have been made to cover the syllabi of nearly all Indian Universities.

The book features both a text and a laboratory guide. It deals with the structures and physiological phenomenon of each system. General topics are also dealt in details. It is hoped that this book will not only meet the requirement of undergraduate students but also be useful as a guideline to the teachers in their teaching. It is also hoped that the general public will find that the book is not riddled with too much of scientific and technical jargon.

It is difficult to acknowledge adequately the assistance of all who have contributed to the preparation of this book. First of all the authors express their thanks to their friends and colleagues whose continuous inspiration have initiated them to bring out the present volume.

The authors are grateful to Dr. J. Bahadur, Professor and Head, School of Studies in Zoology, Jiwaji University Gwalior, Dr. A.K. Pandey, Professor and Head, School of Studied in Zoology, Vikaram University, Ujjain for their valuable suggestions.

Constructure criticism from students and teachers will be highly appreciated by the authors.

H.S. Bhamrah

Kavita Juneja

Contents

1

CHARACTERS & CLASSIFICATION OF MOLLUSCA

Members of the phylum Mollusca are among the most conspicuous invertebrates and include such familiar forms as clams, oysters, squids, octopods and snails. During the last 18th and 19th centuries, when natural history, occupied the time of many well to do gentlemen, shell collections were as popular as today. Such collections, often containing species gathered from various parts of the world, have contributed considerably to our knowledge of the phylum. In abundance of species, molluscs comprise the largest invertebrate phylum aside from the arthropods. Over 100,000 living species have been described. In addition, some 35,000 fossil species are known.

The term Mollusca (L. *mollis* = soft) means soft bodied and for the first time coined by *Aristotle* in his famous *Historia Animalium* to describe various kinds of cephalopods and gastropods. The term Mollusca was used by *Jonston* (1650) to include cephalopods and Barnacles.

The claasification of Mollusca directly into the five classes - Amphineura, Gastropoda, Scaphopoda, Lamellibranchia and Cephalopoda, by *Laseneer* (1992) has remained standard ever since.

Characters.

1. Some are aquatic, mostly marine, a few fresh water and some terrestrial forms.

2. Symmetry bilateral, viscera and shell coiled in Gastropoda and some Cephalopoda.
3. The body is triploblastic made up of 3 germ layers.
4. Body without any appendage and segmentation.
5. Body skin single layered mostly ciliated and with mucous glands.
6. Body usually small and soft and enclosed in an envelop, the mantle that secretes shell of various shape and sizes.
7. Generally the body is divided into 4 parts, the visceral mass carrying the vital organs of the body; the head, the anterior most part bearing mouth, eyes and tentacles but is absent in the *Pelecypoda and* the *Scaphopoda*, the foot a muscular mass, the form and function of which varying in the different classes and lastly the mantle or pallium a thick muscular clock covering the entire animal.
8. Shell, when present, usually univalve or bivalve, consituting an exoskeleton, internal in some.
9. Coelom reduced and represented mainly by pericardial cavity, gonadial cavity and kidney.
10. The alimentary tract is either straight or coiled. Jaws are present in many, especially in gastropods and cephalopods. In moștof animals a rasping organ, the radula ,appears in the buccal cavity. The radula tears up the food.
11. Blood vascular system consists of dorsal heart with 1 or 2 auricles and 1 ventricle and is of open type.
12. Respiration by general body surface or mantle or by one to many gills or a lung in the mantle cavity.
13. Excretion by paired metanephridia (kidneys).
14. Nervous system of paired ganglia, connectives and nerves. Ganglia usually form a circumenteric ring.
15. Sense organs include eyes, statocysts and receptors for touch, smell and taste.
16. Sexes sre usually separate (some hermaphrodite a few protandric); gonads 2 or 1 with ducts; fertilization external or internal. Mostly oviparous, few viviparous.

17. Development direct or interrupted with larval stage, the veliger larva of mollusca. The asexual reproduction is completely absent.

Phylum Mollusca is divided into six classes. Monoplacophora, Amphineura, Gastropoda, Scaphopoda, Pelecypoda and Cephalopoda.

CLASS I MONOPLACOPHORA

1 Abyssal or fossil, from lower Cambrian to Devonian, probably carnivorous.
2. Bilaterally symmetrical and metamerically segmented, but segmentation internal.
3. Shell dorsal and consisting of a single piece (hence Monoplacophora)
4. Mantle covering the dorsal surface of boday.
5. Foot is ventral with a flat creeping sole.
6. Mouth is anterio-median and the anus is posterio-median.
7. Head is without eyes
8. 5 or 6 pairs of gills externally and serially arranged.
9. Six pairs of nephridia and two pairs of gonads.
10. Stomach contains a crystelline style and the animal feeds on radiolarians.
11. Nervous system primitive, lacks ganglia.
12. Sexes are separate.
13. Development is indirect with a trochophore larva.

Example : *Neoplina galatheae.*

CLASS II AMPHINEURA

(Gr; *Amphi* = both; *neuron*= nerve)

1. Most primitive molluscs with either elongated and vermi form or dorso-ventrally flattened body.

2. Head indistinct and without eyes and tentacles.
3. Foot broad, flat, sole-like and ventral in position.
4. Shell and mantle covering the body from dorsal and lateral sides.
5. 6-60 pairs of gills situated laterally in the pallial cavity on either side of foot. They may be reduced to a pair.
6. Mouth and anus at the opposite ends of body.
7. Nervous system primitive and ladder-like with or without ganglia.
8. Heart with one or two auricles.
9. A pair of nephridia connected to the pericardium.
10. Sexes separate or united.

Class Amphineura is divided into two orders :

Order - 1 - Aplacophora

1. The body is long, narrow, worm like.
2. The shell is absent
3. Mantle contains tiny calcareous spicules.
4. Foot is reduced or absent.
5. Heart consists of a single auricle and a ventricle.
6. Gills are absent or reduced to a pair located in the cloacal cavity.
7. Alimentary canal is straight without redula.
8. Nervous system with a distinct brain and ganglia.
9. Sexes united.

Examples : *Chaetoderma, Neomenia*

Order - 2. Polyplacophora

1. Body elliptical, convex dorsally and flattened ventrally.
2. Shell dorsal and formed of eight calcareous pieces arranged in a longitudinal row.

3. Mantle with spicules disposed along the margins.
4. Gills 6 to 8 pairs, present in the pallial groove on the lateral sides of foot.
5. Head devoid of eyes and tentacles
6. The whole ventral surface is occupied by the broad and flat foot.
7. Alimentary canal is coiled with well formed redula.
8. Heart with two auricles.
9. Nervous system without definite ganglia.
10. Sexes separate.

Examples : *Chiton, Chactopleura.*

CLASS - III GASTROPODA

(Gr. *Gastros* = stomach; *podos* = foot)

1. Body is asymmetrical with a mantle and shell.
2. The shell is univalved and spirally coiled.
3. Head is distinct bearing tentacles, eyes, and a mouth.
4. Foot is ventral and muscular
5. Visceral mass usually coiled exhibiting various types of torsions.
6. Anus usually situated anteriosly close to the month.
7. Buccal cavity contains an odontophore with a redula.
8. Respiration is carried on by integument or gills or lungs.
9. Circulatory system open type with a dorsal heart enclosed in pericardium.
10. Excretory system includes one or two kidneys or nephridia.
11. Nervous system contains distinct cerebral, pleural, pedal, parietal and visceral ganglia.
12. Sexes are separate. Mostly oviparous, a few viviparous.
13. Development indirect with veliger and trochophone larval stages.

Class Gastropoda is divided into three sub-classes. Prosobranchia, Opisthobranchia and Pulmonata.

SUBCLAS (A) PROSPBRANCHIA (STREPTONEURA)

1. The visceral nerve commissures are twisted into a figure of 8, hence the name streptoneura (Gr. *streptos* = curved, *neuron* = nerve). This is due to torsion.
2. The mental cavity containing the pallial complex opens anteriorly.
3. The gills or ctenidia, if present, lie in front of the heart hence prosobranchia.
4. Head has a single pair of non-retractile tentacles.
5. Calcarious shell with an operculum.
6. Sexes are separate.
7. Exclusively marine.

Subclass Prosobranchia is divided into three orders : Archaeogastropoda, Mesogastropoda and Neogastropoda.

Order - 1 - Archaeogastropoda (Aspidobranchia)

1. Due to incomplete atrophy of organs, the kidneys, auricles, nephridia and ctenidia are paired.
2. Ctenidia are bipectinate or plume like.
3. Ospheridium poorly developed.
4. The nervous system is little concentrated.
5. Gonad opens into the right kidney serving as gonoduct.
6. Fertilization external.

The order has been distinguished into the following two suborders:

Surborder (i) Docoglossa

1. Shell and visceral mass conical.

2. Ctenidium single or absent or replaced by secondary gills.
3. Radula with three marginals on either side.
4. Auricle single.

Examples :*Patella, Acmaea*

Surborder (ii) Rhipidoglossa

1. Shell spiral.
2. Ctenidia paired.
3. Radula with numerous marginals.

Examples : *Haliotis, Fissurella, Trochus.*

Order - 2. Mesogastrpoda (Pactinibrachia)

1. These are advanced snails with complete atroply of the left side so that usually unpaired auricle, nephridium, ospheridium and ctenidium occur.
2. Ctenidium monopectinate or comb-like, attached to the mantle throughout its entire length.
3. Edge of the shell opening lacks a siphonal notch or canal.
4. Foot may be operculate.
5. Ospheridium pectinate and well differentiated.
6. The nervous system is concentrated.
7. Gonad has its separate duct and opening.
8. Fertilization internal. Development is indirect with a free-swimming veliger larva.

Examples : *Pila, Triton, Valvata, Natica, Vermetus.*

Order - 3. Neogastropoda (Stenopoda)

1. Carnivorous species with an eversible proboscis.
2. Ospheridium is large and bipectinate.

3. Foot operculate.
4. One gill with filaments in one row, heart with one auricle, single nephridium present.
5. Nervous system is lightly concentrated.
6. Shell has a siphonal canal.
7. Embryos are intracapsular. Free swimming veliger supressed.

SUBCLASS (B) OPISTHOBRANCHIA (EUTHNEURA)

1. Due to detorsion of visceral mass, visceral loop becomes untwisted hence named (Gr. *euthus* = straight, *neuron* = nerve).
2. Mantle cavity displaced posteriorly.
3. Head with two pairs of tentacles.
4. Shell and operculum, if present, reduced.
5. Usually one auricle and one nephridium present.
6. Ctenidia are replaced by secondary gills.
7. Hermaphroditism is universal. Larva is veliger.
8. Exclusively marine.

Subclass Opisthobranchia is divided into eight orders as follows:

Order - 1. Cephalaspidea

1. These are burrowing forms having shield like head.
2. Shell and mantle cavity moderatley developed.
3. Lateral parapodial lobes are prominent.

Examples : *Bulla, Acteon.*

Order - 2. Anaspidea

1. These are crawling orburrowing forms having a pair of rhinophores on head.
2. Shell is small and internal.

3. Mantle cavity is reduced on right side.
4. Parapodial lobes are prominent.

Examples : *Aplysia, Akera.*

Order - 3. Thecostomata

1. These are shelled pteropods.
2. Shell is spirally coiled or a non-spirally pseudo-conch.
3. Mantle cavity is well developed.
4. Parapodial fins are large.

Examples : *Clio, Limacina*

Order -4. Gymnostomata

1. These are naked pteropods and planktonic.
2. Shell and mantle cavity absent.
3. Small ventral parapodial fins are present.

Example : Pneumoderma.

Order - 5. Notaspidea

1. Shell is external or reduced and internal.
2. Mantle cavity is absent, but a skirt like projection of mantle covers gill on the right side.

Example : *Pleurobranchus.*

Order - 6. Acochlidiacea

1. Small without shell and gill.
2. Naked visceral mass projecting behind the foot and covered with spiracles.

Examples : *Acochlidium.*

Order. 7. Saccoglossa

1. These are herbivorous, with a modified redula and suctorial pharynx.
2. They are shelled or naked slug-like, with a gill.

Examples : *Elysia, Oxynoe.*

Order - 8. Nudibrandia (Acoela)

1. They are naked bilaterally symmetrical sea-slugs without shell, mantle cavity, ctenida and osphridium.
2. They have secondary gills around anus or susface outgrowths or cerata.

Examples : *Doris, Aeolis, Tritonia.*

SUB CLASS - C- PULMONATA

1. The ctenidia are absent and replaced by pulmonary sac or lung.
2. Shell is simple spiral or vestigial or absent and operculum never occur.
3. One or two pairs of tentacles and one pair of eyes are present.
4. The torsion takes place. The nervous system is secondarily symmetrical due to shortening of connectives and concentration of ganglia into a circumoesophageal ganglionic complex.
5. The heart with a single auricle lying anterior to the ventricle.
6. The animals are bisexual. Gonad single.
7. Development direct without larval stage.

The subclass is divided into two orders:

Order - 1 Basommatophora

1. One pair of tentacles with the eyes at the base.
2. Shell is delicate with a large aperture.
3. Some forms with secondary gills.

4. Male and female gonopores usually separate.

Examples : *Limnea, Planorbis, Physa.*

Order - 2. Stylommatophora

1. Two pairs of retractile tentacles. Eyes are lodged on the tip of posterior pair of tentacle.
2. Shell is internal, reduced or absent.
3. Mostly terrestrial.
4. Male and female gonopores usually united.

Éxamples : *Helix, Limax, Arion*

CLASS. IV. PELECYPODA

1. The pelecypods or bivalvia are marine, fresh-water, bilaterally symmetrical molluscs.
2. Mantle consists of paired right and left lobes, which secrete a bivalved shell.
3. Devoid of head, eyes, tentacles and redula.
4. The muscular foot is usually wedge-shaped or tongue-shaped, which is adapted for burrowing.
5. Presence of two gills, one on either side of mantle cavity, the function of which is to produce respiratory and food-carrying currents of water.
6. Pharynx is absent.
7. Coclom is reduced to a dorsal placed pericardium.
8. Alimentary canal is coiled with large paired digestive glands.
9. Heart is continued with pericardium and comprises two auricles and one ventricle.
10. The renal organs are a pair of coelomic kidneys, which lead into pericardium at one end and to outside at other.
11. The nervous system includes four pair of ganglion, *viz.*, cerebral. pleural, pedal and visceral.

12. Sense organs are statocyst and osphradium.

13. Sexes may be separate or united. Development includes metamorphosis, followed by trochophore stage.

Class - Pelecypoda is divided into following orders:

Order. 1. Protobranchiata

1. Presence of paired plume-shaped gills, each bears two rows of filaments.
2. The foot is flattened.
3. Presence of anterior and posterior adductor muscles.
4. Sexes separate, gonads lead into kidneys.

Examples - *Nucula, Solenomya*

Order. 2. Filibranchiata

1. Single pair of plate-like gills formed of distinct V-shaped filaments.
2. Inter-filamentar junctions are either absent or formed by groups of inter-locking cilia.
3. Inter- lamellar junctions are either absent or non-vascular.
4. Two adductor muscles present, anterior may be redused or absent.
5. Foot small or poorly developed.

Example : *Mytilus, Arca, Noah.*

Order . 3. Pseudolamellibranchiata

1. The gills are plaited so as to present vertical folds.
2. Interfilamentar junctions may be ciliary or vascular. Interlamellar junctions may be vascular or non- vascular.
3. Shell valves are frequently unequal and single large adductor muscle present.
4. Foot is little developed or rudimentary.

5. Genital organs open into kidneys or near them.

Examples : *Pecten, Ostrea, Pinna, Lima.*

Order - 4 Eulamellibranchiata

1. Gills are basket like, the gill-filaments are attached with vascular inter-filamentar and inter-lamellar junction.
2. Gills may be smooth or plaited.
3. Two equal adductor muscles.
4. Foot is well developed.
5. Siphon of small of large sized present.
6. Gonads open out by separate openings. They do not open into kidneys.

Examples : *Unio, Teredo, Anodonta, Venus, Cardium, Mya.*

Order - 5. Septobranchiata

1. Gills replaced by horizontal muscular and perforated septa dividing the mantle cavity into upper and lower chambers.
2. Foot long and slender, byssus reduced or absent.
3. Adductor muscles two.
4. True ctenidia are absent.

Examples : *Cupisdaria, Poromya.*

Order. V. Scaphopoda

1. They are aberrant marine animals comprising only of three genera.
2. Body is elongated worm-like,enclosed in a tusk-like shell open at both ends.
3, Eyes, tentacles and gills are absent.
4. Mantle tubular completely enclosing the body.
5. Mouth surrounded by lobular processes or outgrowths.

6. Foot is reduced used for digging.
7. Heart rudimentary.
8. Sexes are separate.
9. Torsion takes place in some forms

Scaphopoda has two families :

Family (i) Dentalidae

1. Foot is trilobed having two epipodial lobes and one hypopodial lobe.

Examples: *Dentalium, Antalis.*

Family (ii) Siphonodentalidae

1. Foot is elongated and capable of expansion into a terminal disc.

Examples : *Pulsellum, Siphonodentalium.*

CLASS - VI CEPHALOPODA

1. Exclusively marine.
2. Shell spiral chambered or usually with or without shell embedded in mantle.
3. Body with head and trunk. Head bears eyes and mouth.
5. Trunk consists of symmetrical and uncoiled visceral mass.
6. Mantle encloses posteriorly and ventrally a large mantle cavity.
7. Foot altered into a series of sucker bearing arms of tentacles encircling the mouth.
8. Mouth bears jaws and radula.
9. Two or four pairs of bipactinate gills.
10. Circulatory system closed, heart with two or four auricles.
11. Excretory system comprises two or four pairs of nephridia.
12. Nervous system is highly developed and the principal ganglia are concentrated around the oesophagus.
13. Sexes are separate.
14. Development meroblastic without metamorphosis
15. Most of the cephalopods have an ink-gland with a duct opening into the rectum.

Cephalopoda is divided into three subclasses:

SUBCLASS - A. DIBRANCHIATA

1. The foot modifies into a circlet of eight to ten arms bearing suckers, around the mouth.
2. The funnnel forms a complete tube.
3. The shell is generally internal.
4. Presence of paired gills and kidneys.
5. Presence of characteristic ink-gland and duct.

Dibranchiata is divided into two orders

Order (1) Octopoda

1. Body rounded or oval and with no lateral fins.
2. Arms eight, long arms or tentacles absent.
3. Suckers on the arms sessile and without horny rim.
4. Shell and cnidamental glands absent.

Examples. *Octopus* (Devil-fish), *Agronauta*

Order (2) Decapoda

1. Body elongated and with no lateral fins.
2. Arms ten, eight short and 2 long and known as tentacles.
3. Suckers on the arms sessile and without horny rim.
4. Shell internal and well developed.

Examples : *Sepia, Loligo,*

SUBCLASS - B. NAUTILOIDEA OR TETRABRANCHIATA

1. The foot divides into lobes bearing numerous tentacles.
2. The funnel is incomplete.
3. Shell is external which is divided into several compartments.
4. Kidneys and gills are four.

5. Devoid of ink gland.
6. Pericardium communicates to the exterior directly.
7. Eyes simple without crystelline lens.
8. Chrometophore, salivary glands absent.

Example : *Nautilus*.

SUBCLASS - C . AMMONOIDEA.

1. All are extinct.
2. Shell variously modified in the form of straight spiral or turreted.
3. Siphon external, siphuncle simple and marginal.
4. Two pairs of ctenidia.

Example : *Ammonites*

CHAETODERMA

Phylum	-	Mollusca
Class	-	Amphineura
Subclass	-	Aplacophora
Genus	-	*Chaetoderma*

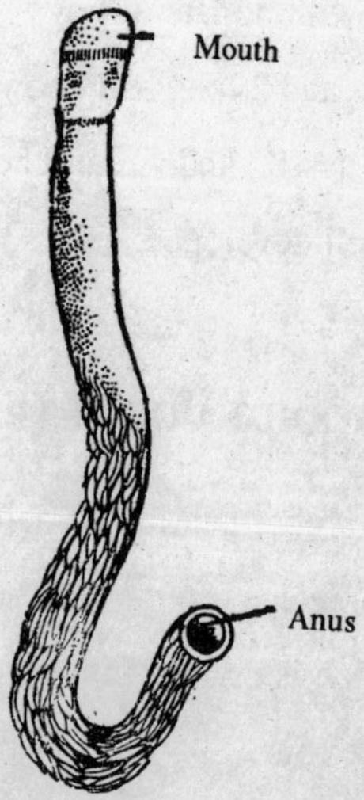

Fig. 1.1 *Chaetoderma*

1. It is a most primitive mollusc. The body is wormlike cylindrical. It is found around North Atlantic.
2. The head is separate off from the body by a constriction.
3. Mantle covers the body completely.
4. Shell is absent but the body is covered by numerous calcareous spicules.
5. Rudimentary foot.
6. Gills reduced to a pair which are situated in the cloacal cavity.
7. Sexes are separated and each has one fused gonad.
8. A cavity is present at the posterior end of the body into which opens the anus.

PATELLA

Phylum	-	Molluca
Class	-	Gastropoda.
Sub Class	-	Prosobranchia
Order	-	Aspidobranchia
Suborder	-	Docoglossa
Genus	-	*Patella.*

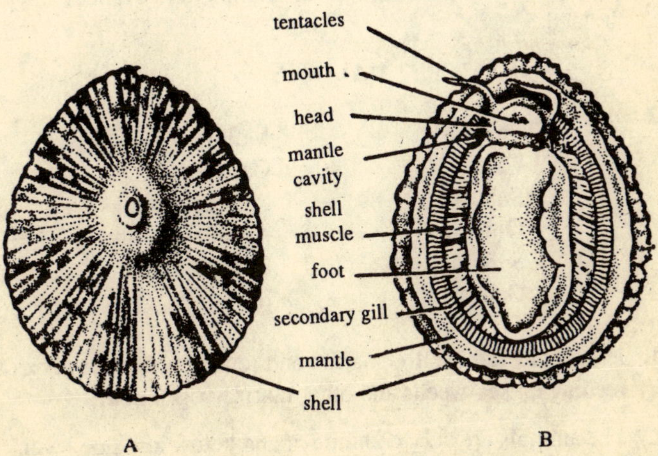

Fig. 1.2 *Patella*. A -- Dorsal view; B -- Ventral view

1. *Patella* or limpet is a littorral and is found attached to the rocks.
2. It is found along Pacific coast, Atlantic coasyt, Europe, U.S.A.
3. Body is oval with convex dorsal surface.
4. Shell is oval and core - like, with a conical projection from its dorsal surface.
5. Operculum is absent.
6. The head is laterally produced into a pair of small, stout, tectile tentacles.
7. Eyes are simple present at the base of tentacles.
8. Foot ventral and sole-like for creeping.
9. True mantle is restricted to the anterior and and secondary mantle cavity is developed between foot and mantle.
10. Redula composed of very few, strong hooked teeth in each row.
11. All around the foot lies a secondary pallial cavity containing a series of secondary branchiae•
12. Heart with single auricle.
13. Herbivorous, feeding upon algae, sea-weeds and diatoms.
14. It is eatern by poor classes in country like France, Italy and Ireland.

HALOTIS

Phylum	-	Mollusca
Class	-	Gastropoda
Subclass	-	Prosobranchia
Order	-	Aspidobranchia
Suborder	-	Rhipidoglossa.
Genus	-	*Halotis*

1. *Haliotis* or Ear shell is marine and found attached to rocks and feeding on sea-weeds and other marine vegetation.
2. The animals are very common at and below low tide level.
3. *Haliotis* or *Abalone* has ear-shaped shell with small flattened spire.

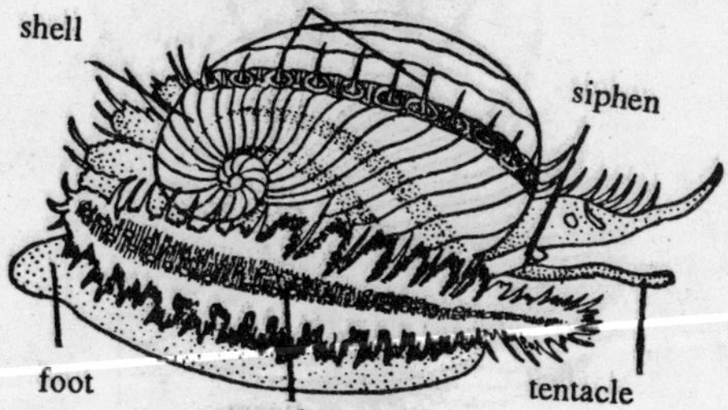

Fig. 1.3. *Haliotis*

4. The aperture is very large and without operculum.
5. The shell is perforated by a series of marginal slits through which project the tentacular processes of the mantle.
6. The foot is large and with epipodia and the animal can move 2 to 4 metres per minute but clings very powerfully to the boulders.
7. Eyes are borne on stalks at the outer bases of the tentacles.
8. The mantle cavity is large, has bipectinate ctenidia or gill, the right being smaller.
9. The shell of the animal is of great economic importance, as beautiful pearly buttons are manufactures from the shell of *Haliotis* in Europe and U.S.A.
10. The animal is also eaten in countries like China, Japan and many cities along the Pacific coast.

MUREX

Phylum	-	Mollusca
Class	-	Gastropoda
Subclass	-	Prosobranchia
Order	-	Pectinibranchia
Genus	-	*Murex*

1. It is a marine animal found between low tide marks on stones and sea weeds.

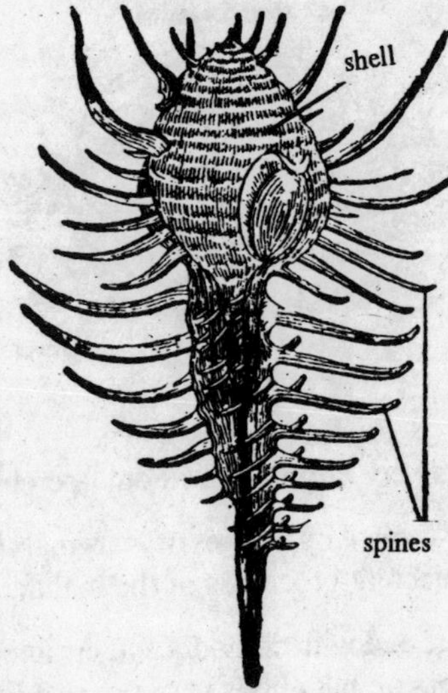

Fig. 1.4 *Murex.*

2. It is found on the Syrian coast, Italy, India, West Indies and U.S.A.
3. Shell is spirally coiled and with three or more rows of beautiful spines of different lengths.
4. It is carnivorous and highly predator feeding on living animals which it grasps by means of its foot. It is destructive to oyster bed.
5. A very prominant bipectinate osphradium persist.
6. Head is prolonged into a long proboscis which can be retracted within the probocis-sheath.
7. Adrectal glands present which secrete a dye the *Jyrain purple* (Yellow in colour).
8. Sexes are separate.

APLYSIA

Phylum	-	Mollusca
Class	-	Gastropoda
Subclass	-	Opisthobranchia

Order - Anaspidea
Genus - *Aplysia.*

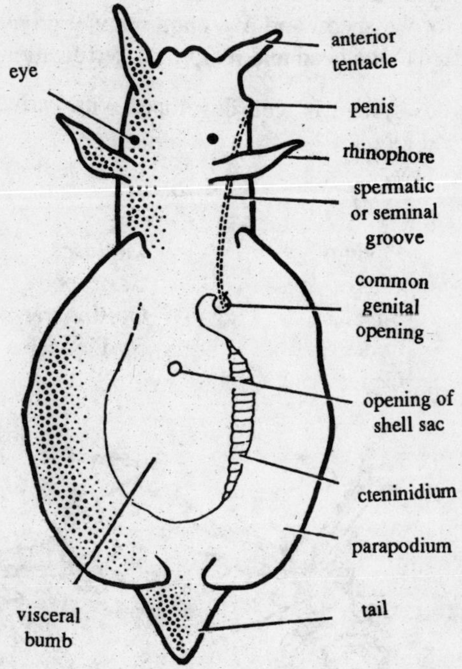

Fig. 1.1 *Aplysia.*

1. It is a marine animal found among sea -weeds.
2. It is found in India, Asia, West Indies and on the Florida coast.
3. It is commoly called 'sea hare'.
4. Body is soft, lumpy with a visceral hump.
5. Head with two pairs of tentacles. The anterior pair is large and ear-like, the posterior pair is olfactory and bear eyes at their base.
6. Shell thin, reduced and transparent, enclosed by mantle.
7. The foot is muscular and elongated pointed posteriorly with lateral out growths, the parapodia.

8. Mantle possesses unicellular ink glands which secrete purple ink, used for defence.
9. Hermaphrodite. There is a single generative aperture and a single duct for the sperm and ova but a seminal groove runs from the aperture to the head and reciprocal fertilization is impossible.
10. Nervous system is well developed with perfectly symmetrical visceral loop.

EOLIS

Phylum	-	Mollusca
Class	-	Gastropoda
Subclass	-	Opisthobranchia
Order	-	Nudibranchia
Genus	-	*Eolis*

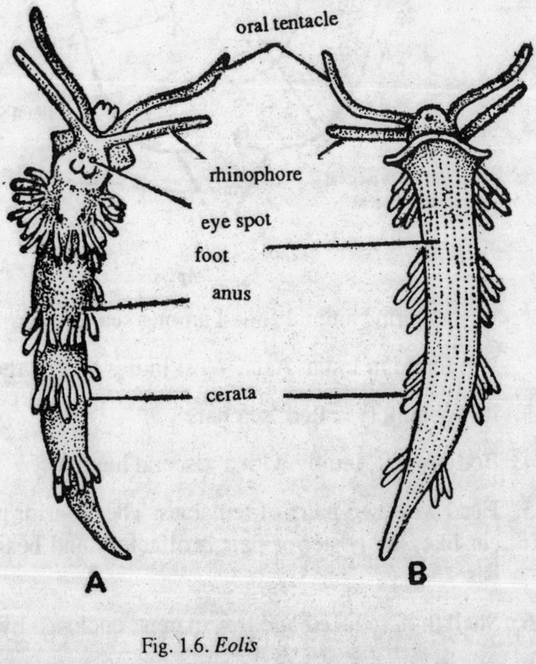

Fig. 1.6. *Eolis*

1. It is a marine slug- like gastropod found in shallow waters crawling on under surface of sea weeds.

2. Found in U.S.A., Europe.
3. *Aeolis* is a small nudibranch 1 or 2 cm in length.
4. The soft body is covered with a transparent integument.
5. The shell, mantle and true gills are absent.
6. The cerata are the respiratory organs.
7. There are cylindrical extensions of the integument, containing hollow extensions of the liver and nematocysts derived from ingested hydroids.
8. The cerata have characteristic arrangment, may occur in many transverse or in few longitudinal rows.
9. The head is small, bears tentacles, usually two pairs, one of which may be retractile.
10. The anterior or oral pair is the longer.
11. The posterior or dorsal pair, known as rhinophores bear simple sense organ which test the surrounding water.
12. At the base of of each rhinophore is an eyespot.
13. The mouth is prominent and can be most easily seen from the ventral view.
14. On the right side of the head is conspicuous genital opening.
15. Just posterior and dorsal to genital opening lies the small anus.
16. On the ventral side the animal has a highly muscular foot.
17. Proportion varies with the species.
18. Mouth is guarded by lips which cover both the radula and the horny jaws within the buccal cavity.
19. The powerful jaws cut the hydroid into two pieces. The *Aeolis* is hermaphroditic and gonad is protandric.

DORIS

The systematic position is same as that of *Aeolis*

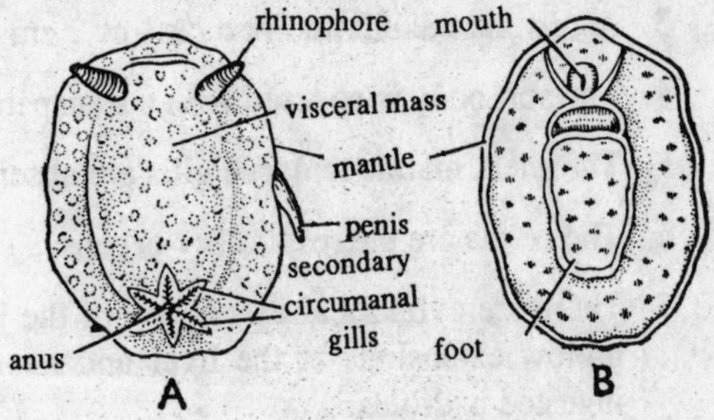

Fig. 1.7 *Doris*

1. It is a sluggish marine animal found under stones at low tide marks.
2. It is cosmopolitan in distribution.
3. It is commonly known as 'sea lemon'.
4. The body is oval and depressed, concave abopve and flat below.
5. Head is indistnict, but there is a pair of short olfactory tentacles or rhinophores towards the anterior end.
6. Mantle is usually pigmented and contains calcareous spicules or dorsal tubercles.
7. Mouth is present on ventral side. Anus lies mid-posteriorly and is surrounded by a circlet of feathered retractile secondary branchiae.
8. The broad foot is on ventral side.
9. Hermaphrodite, gonopore is asymmetricelly placed on the right side of the body.
10. *Doris* has undergone *detorsion*, therefore, the shell, mantle cavity and primary gills are absent.

LIMAX

Phylum	-	Mollusca
Class	-	Gastropoda
Subclass	-	Pulmonata

Order - Stylommatophora
Genus - *Limax.*

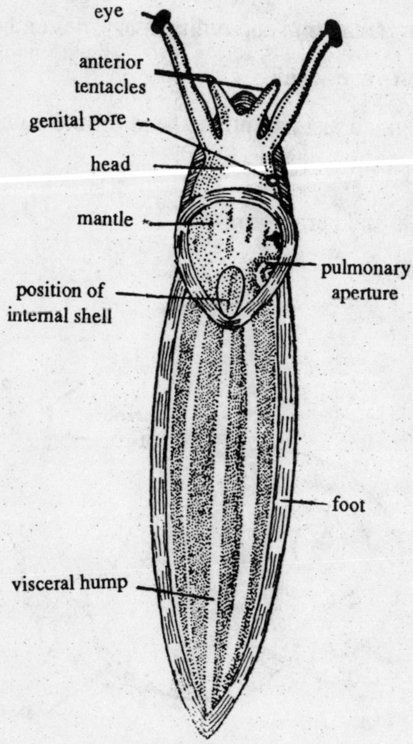

Fig. 1.8 *Limax:*

1. It is a terresttial gastropod, found in gardens over damp soil.
2. It is cosmopolitan in distribution.
3. It is commonly known as 'gray slug'.
4. Body is elongated and tapering behind and is divisible into head, foot and visceral hump.
5. Two pairs of retractile tentacles are present on head. Eyes are present at the tips of posterior tentacles.
6. The mouth is on the ventral side of the enterior end and is bounded by a pair of lateral lips.

7. Mantle forms a shield like area in the dorsal side of the anterior part of body. The mantle cavity acts as a lung for the breathing, pneumostome, present on right side.
8. Shell is internal, thin and rudimentary, lies embedded in mantle.
9. Foot is on ventral side.
10. A pedal gland just behind the head for the mucous secretion.
11. Bisexual.
12. Nocturnal and herbivorous.

HELIX

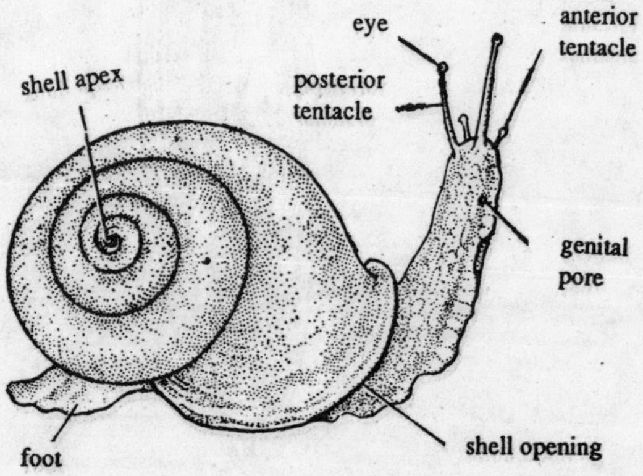

Fig. 1.9. *Helix*

The systematic position is same as that of *Limax*

1. It is a terrestrial animal found in damp places, under bark and fallen leaves.
2. It is found in palearctic region.
3. It is commonly called 'garden snail'
4. The shell is thin, external and heliciform with a low conical spire, measuring 4 cm. in length.
5. Head bears two pairs of tentacles, first pair bears organs of small and second pair is large and bears a pair of simple eyes.

6. Foot possesses a flat ventral surface for creeping.
7. Respiration by pulmonary sac.
8. Gills are absent.
9. Hermaphrodite. The genital aperture opens above the right lateral lip. Gonads contain four angled dart and multifid digitiform glands.

MYTILUS

Phylum	-	Mollusca
Class	-	Pelecypoda
Order	-	Fillibranchia
Genus	-	*Mytilus*

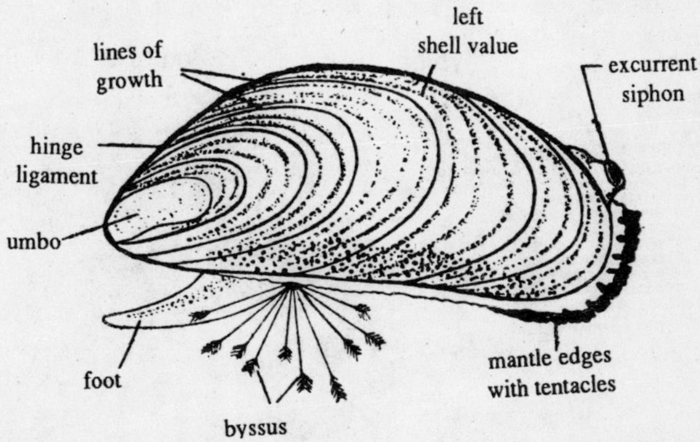

Fig. 1.10 *Mytilus*

1. It is a marine, sedentary mollusc found at a depth of 2 or 3 fathoms in low tide mark, attached to rocks or wooden structure by its byssus threads.
2. It is commonly called as 'sea mussel' and cosmopolitan in distribution.
3. The shell is elongated, equivalent with umbo at or near anterior end.

4. Hinge toothless but may bear cerculations.
5. The foot is tongue like and is modified into a bunch of byssal threads coming out from between the two shell- valves, which are main anchoring organs.
6. The anterior adductor muscle is weaker where as posterior is strongly developed.
7. A pair of gills is present, each gill is provided with gill-filaments. The gills are lamelliform.
8. The eyes are found anterior to the inner gill-lamella.
9. Sexes are separate. Gonads extend into the mantle.
10. It is used as food in Europe.

PECTEN

Phylum	-	Mollusca
Class	-	Pelecypoda
Order	-	Pseudolamellibranchia
Genus	-	*Pecten*

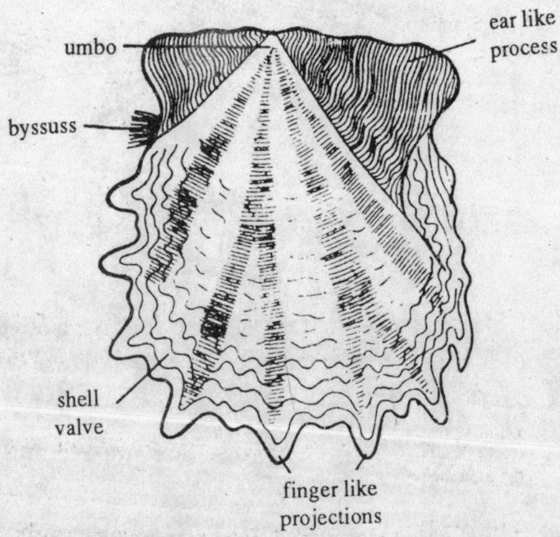

Fig. 1.11. *Pecten.*

1. It is free swimming marine molluse lives on the sea bottom of 10 fathoms deep.
2. It is found in India and U.S.A.
3. It is commonly called as 'scallop'.
4. Two shells are unequal. The right shell being larger and more convex and the animal rests on this valve.
5. There is a single large adductor muscle. This is divided into parts and the larger of these served for rapid contraction which cause swimming movements.
6. Foot is very much reduced but in larva very well developed and used for locomotion.
7. Two large and crescentric gills are present.
8. Mantle is *tentaculiferous* and it encloses viscra.
9. Large number of stalked eyes are present at regular intervals along the edges of the mantle.
10. These are hermaphrodite.
11. The ovary has pink colour when eggs are ripe.
12. The testis lies behind and is cream in colour.

OSTREA

The systematic position is same as that of *Pecten*.

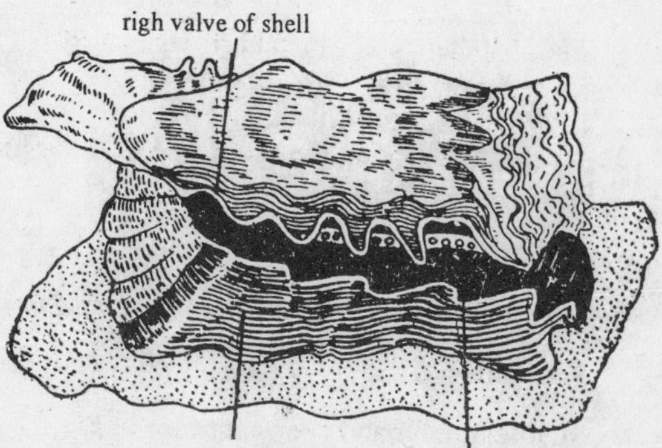

Fig. 1.12. *Ostrea*

1. It is a sedentary bivalve attached to rocks or other shells and found in shallow and brackish water.

2. It is found in Atlantic and Pacific coasts, India, Gulg of Mexico to Massachusetts.

3. It is commonly called 'edible oyster'.

4. The shell valves are unequal, the left is always larger and permanently attached to the rocks by byssus.

4. Ony one adductor muscle i.e. posterior one. It is divided into two parts one is striated and other is non-striated.

5. The foot is absent.

6. Heart with two fused auricles.

7. A pair of curves gills present.

8. It is peculiar in the sense as the individuals function alternately as males and females.

9. Spawning generally takes place at full moon. The eggs develop into vileger larva which metamorphoses into adult.

TEREDO

Phylum	-	Mollusca
Class	-	Pelecypoda
Order	-	Eulamellibranchia
Genus	-	*Teredo*

1. Marine pelecypod which burrows in wood of the boats, ships and other submerged wood.

2. It is found in India, Europe, Cape Cod to Florida and Massachusetts Bay to Florida.

3. It is commonly called 'shipworm.

4. *Teredo* has body extremelly elongate and worm-like appearance.

5. Foot reduced.

6. Mantle tubular and opens anteriorly, shell is reduced and covers only the anterior part of the body.

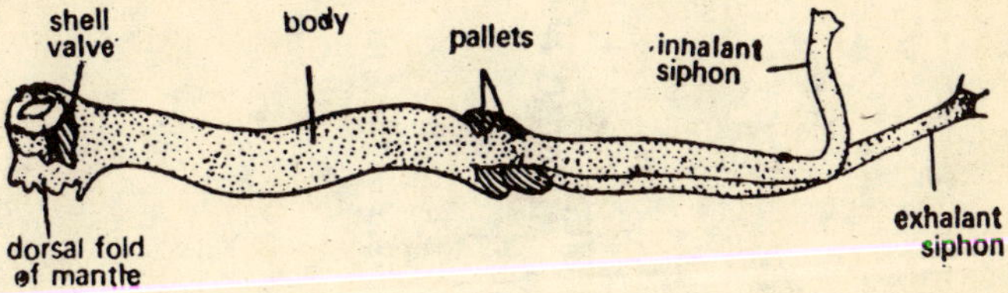

Fig. 1.13 *Teredo*

7. The mantle cavity and siphons are long and the latter unites to a large extend and provided with calcareous pallets for protection.
8. A constant current into and out of the mantle cavity is maintained by ciliary action.
9. The saw dust is swallowed by the animal and is digested by cellulose digesting enzymes.

SOLENOCURTUS

Phylum	-	Mollusca
Class	-	Pelecypoda
Order	-	Eulamellibranchia
Genus	-	*Solenocurtus.*

1. It is a burrowing pelecypod found in India and U.S.A.
2. It is commonly called 'razor fish' or 'razor shell'
3. Bivalve shell, mantle divides into right and left lobes.
4. Body is laterally compressed and bilaterally symmetrical.
5. A pair of basket-shaped gills consisting of filaments, which are united by interfilamentar and inter -lamellar junctions.
6. Two equal-sized adductor muscles developed.
7. Presence of two elongated siphonal tubes which are united with each other.

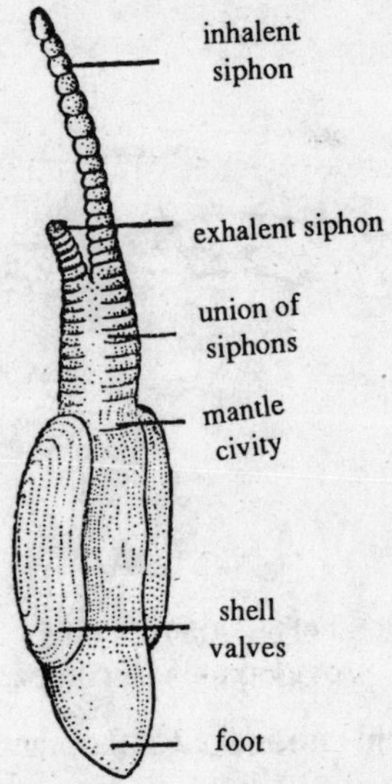

Fig. 1.14. *Solenocurtus* (Razor-fish)

8. The shell has the appearance of long scabbard.

9. The foot is long, cylindrical and enormously developed which is capable of penterating into the loose sand.

OCTOPUS

Phylum	-	Mollusca
Class	-	Cephalopoda.
Subclass	-	Dibranchia
Order	-	Octopoda
Genus	-	*Octopus*

1. It is marine and cosmopolitan in distribution generally found on the Atlantic and Pacific coasts.

2. It is commonly called devil fish.

3. Body surrounded by mantle containing chromatophores. They can change their colour according to surrounding medium.

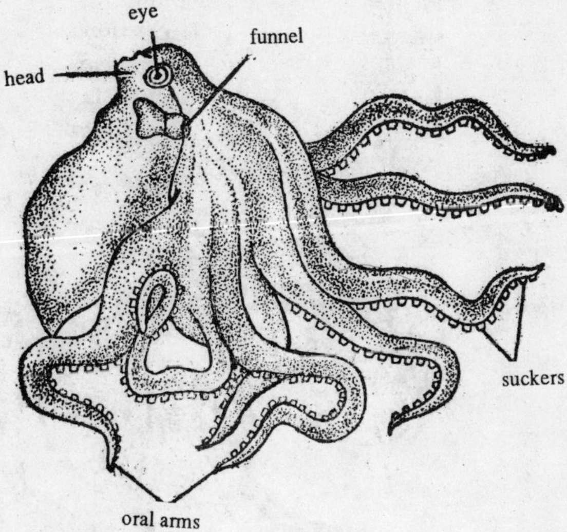

Fig. 1.15 *Octopus*

4. Foot modified into eight equal arms surrounding the head and siphon.

5. The arms are beset with suckers which are sessile and devoid of horny rims.

6. Body is covered with tubercles, short, round aborally and without fins.

7. External shell is absent and body with vestiges of an internal shell.

8. Third arm on the right in male modified to act intermittent organ or hectocotylized.

9. Like *Sepia and Loligo, oitopus* is also capable of producing inky liquids which form a smoky screen water.

10. Sexes are separate and there is no metamorphosis as the development is direct. Female cares the eggs.

11. Mouth with sharp and beak-like horny jaws

ARGONAUTA

Phylum	-	Mollusca
Class	-	Cephalopoda
Subclass	-	Dibranchia
Order	-	Octopoda.
Genus	-	*Argonauta.*

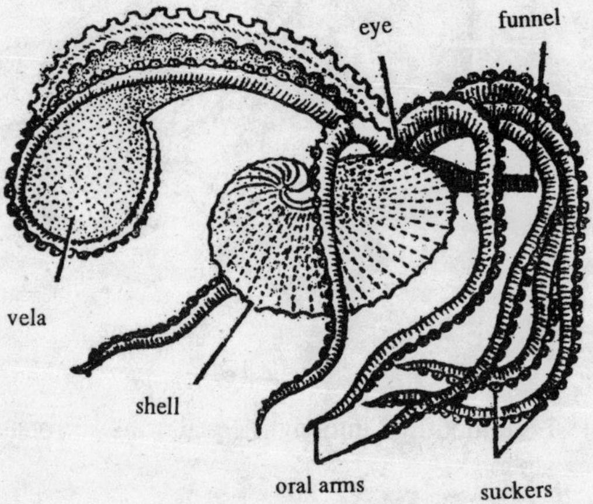

Fig. 1.16 *Argonauta*

1. It is a marine animal found in Atlantic and Indian oceans.
2. It is commonly called as 'paper nautilus'.
3. Sexual dimorphism is remarkable and well-marked.
4. Male in 2.5 cm in length and without shell.
5. Female is 20 cm in length and possesses a thin, transparent shell secreted by the extrimities of two dorsal arms.
6. Shell is used for protection of eggs.
7. Third arm in male is hectocotylized.
8. Eyes and funnel and present.
9. Two plume like ctenidia and two kidneys are present.

NAUTILUS

Phylum	-	Mollusca
Class	-	Cephalopoda
Subclass	-	Tetrabranchia
Genus	-	*Nautilus*.

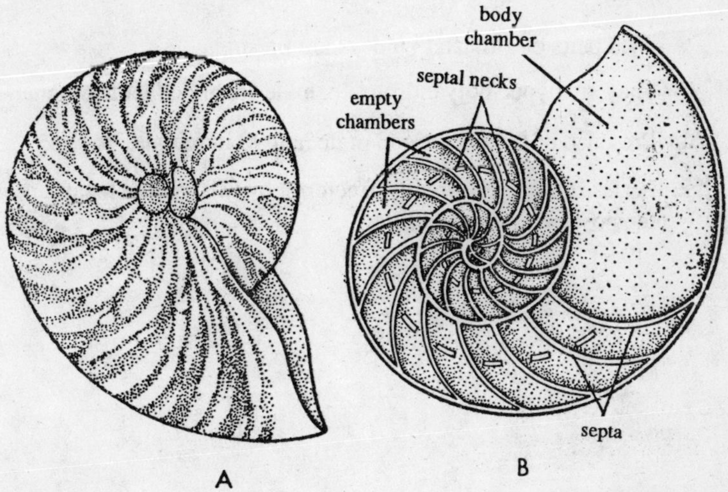

Fig. 1.17 *Nautilus* (A) External Shell (B) Internal Chamber

1. It a marine cephalopod living in shallow waters and shores and coral reefs in the Indian and south Pacific Oceans.
2. It is commonly knwon as 'Pearly nautilus'.
3. The shell is well developed, spirally coiled and composed of calcareous matter, which is divided into several compartments by a series of septa.
4. Through the various compartments runs a tube called siphuncle the walls of which are formed of calcium carbonate. A narrow vascular prolongation of visceral region of the animal passes through this tube.

5. Head is well developed bearing the eyes.
6. The anterior part of foot is modifies into several arms provided with tentacles, which are arranged in two series.
7. The posterior portion of the foot is modified into funnel situated below the head.
8. Buccal cavity with a pair of jaws and odontophore bearing several teeth.
9. Two pairs of gills and kidneys are presnt.

10 They are hyper-polyandrous i.e. male are abundant than female.

11. Presence of 60 tentacles in male and 90 in female.
12. Pericardium leads directly to ecterior by two orifices located near the posterior renal aperture.

2

CHITON

Although some features of their structure and embryology are primitive, Chitons have become highly adapted for adhering to rocks and shells. Their body is characterized by a large foot occupying the whole of ventral face of the body, with eight calcareous plates on the mantle and a complete row of branchiae between foot and mantle. These are included in class Amphineura due to the presence of longitudinal nerve cords on both the sides of the shell (*Amphi* = both, *neura* = nerva).

SYSTEMATIC POSITION

Phylum	-	Mollusca
Class	-	Amphineura
Order	-	Polyplacophora
Genus	-	Chiton.

Distribution

Chitons are popularly known as 'sea mice' or 'Coat of mail shell' are all marine. They are found all over the world at all the depths, though most abundant on the shores between tidal limits. There are approximately 600 existing species of Chitons. The fossil record, which dates back to the upper Cambrian Period, is rather sparse. Some 350 fossil Chitons are known.

Habits and Habitat

The chitons are mostly nocturnal, shy, and sluggish animals. They prefer dark areas during day time but creeping about the rocks at night to feed on diatoms or small encrusting algae, scraped by the use of radula. Thus they are herbivorous. When at rest, they adhere firmly to the under surface of stones by pressing their powerful sucker like foot and ventral surface of the mantle against the object. When disturbed, they clamp the foot and mantle still more firmly and are then almost impossible to remove without injury. They creep slowly, but when dislodged, they promptly roll up like a pill bug or millipede, perhaps a defensive mechanism.

External features

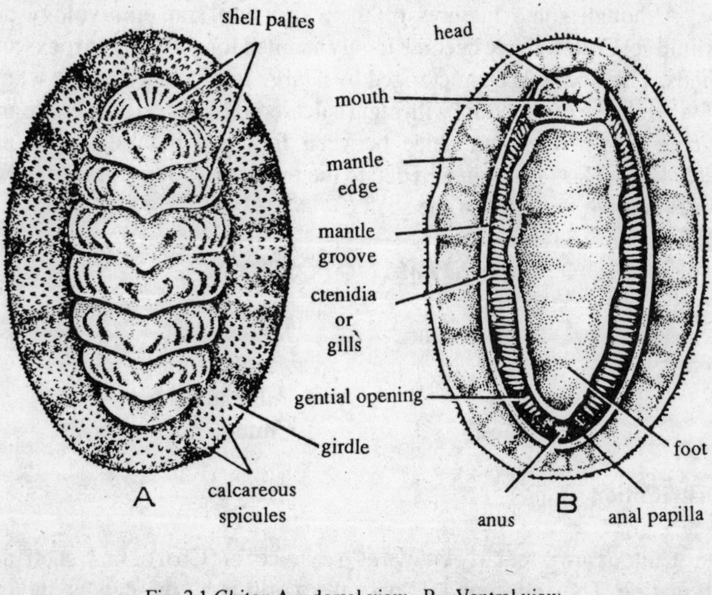

Fig. 2.1 *Chiton* A = dorsal view, B-- Ventral view.

Shape. The body is elongate, elliptical with a convex dorsal surface bearing eight overlapping calcareous plates protective in nature. On ventral side is foot, head and visceral mass.

Size. Chiton species range in size from a few centimeters or less. *Chaetopleura apiculata* is 1 cm. in length found in Atlantic Coast. The giant Pacific species, *Cryptochiton stelleri* measures 30 cm. in length. However, most species are 3-12 cm. in length.

Colouration. Chitons are usually not brighly coloured; dark shades of red, brown, yellow and green are common.

Shell and Mantle. The most distinctive characteristic of chitons is the shell. The shell is divided into eight overlapping transverse plates. From the nature of the shell is derived the name of the order Polyplacophora. Each plate is similar to the others except the first and last. The first is called *cephalic plate* and the last is *anal plate* which are hemispherical in shape. Except for the overlapping posterior edge, the margins of each plate are covered by mantle tissue. Thus the shell is partially embedded in mantle.

The shell plates are composed of several layers. The upper layer, called the *tegumentum*, consists of an organic conchiolin matrix impergnated with $CaCO_3$ and is sculptured on its exposed surface. It is perforated by numerous vertical canals containing sense organs. The tegumentum is covered by a thin *periostracum*. Beneath the tegumentum lies a thicker and denser layer, called the *hypostracum*. The hypostracum is made up of $CaCO_3$. The anterior articulating projections of each plate are composed only of the hypostracum, which in these specialized areas of the plate is called the *articulamentum*.

The shell plates are movable upon one another and allow the animal to roll up like a wood-louse.

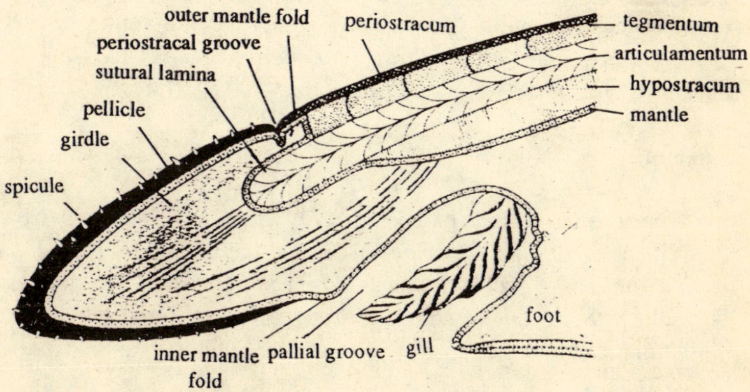

Fig. 2.2. *Chiton*, T.S. through gidle, mantle groove and foot:

The peripheral area of mantle, called the *girdle*, is very heavy and extends a considerable distance beyond the lateral margins of the plates.

The girdle surface displays a variety of ornamentation. It may be naked and smooth or covered with scales, bristles or calcareous spicules. The bristles or spicules may be so long and dense that the animals has a shaggy appearance.

Foot and Locomotion The foot is broad, flat and muscular powerful organ present on the ventral side of the body. It functions in adhesion as well as in locomotion. Adhesion is effected by both the foot and the girdle. The foot is responsible for ordinary adhesion as it is abundantly supplied with a slimy secretion but when it is disturbed, the girdle is also employed.

Chitons are very sluggish animals and may remain in one area for a long period. Most species are motionless at low tide. It moves by employing mucus and waves of muscular activity.

Head. The head is small, cylindrical, inconspicuous structure and is not distinctly marked off from the rest of the body. It contains the month mid-ventrally. The eyes and tentacles are absent. The head is separated from the foot by a narrow groove. The head is provided with a labial palp on either side of the month.

Mantle cavity and groove. A space is left in the form of shallow groove between mantle and foot and is called *pallial groove*. The ctenidia or gills are lodged in this groove being arranged in a single row on each side. The pallial groove also contains the mid-ventral *anus*, two *renal apertures* one on each side slightly in front of anus and two *genital apertures* one on each side in front of the renal apertures.

Coelom.

The body cavity is an extensive cavity and is divisible into two parts, the primary body cavity or *haemocoel* and the secondary or *true body cavity*. The haemo coel forms an extensive system of lacunae or sinuses, into which the arteries open. The true coelom is lined by coelomic epithelium and is represented by the pericardium and genital cavities.

Digestive System

The alimentary canal is simple, extends from one end of the body to the other. It comprises:

Mouth. The mouth is situated mid-ventrally on the head leads into a small buccal cavity.

Buccal cavity. It is small, cylindrical and lined with chitin. The lining of buccal cavity is thrown into longitudinal folds. The floor of buccal cavity contains a muscular elevation, the *odontophore*, with a long, horny and ribbon like *redula*. The redula bears 17 teeth in each transverse row. The lateral teeth are capped with magnetite (iron). The redula is contained in a *redular sac* above the odontophore. Below the odontophore is a smaller evagination called the *subredula sac*. Subredula sac contains a cushion-shaped sensory structure, the *subredula organ*, hanging from the roof. A pair of salivary glands opens through the anterior dorsal wall of the buccal cavity. Periodically the subredula organ is protruded and tests the substratum again.

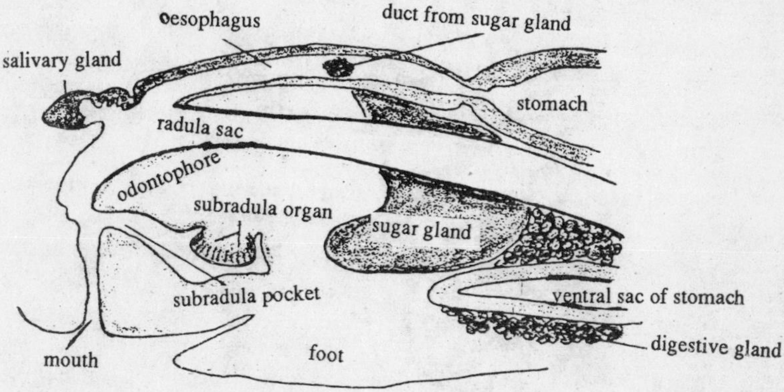

Fig. 2.3. *Chiton*, Buccal region

Oesophagus. The buccal cavity opens into a short oesophagus which is lined with cilia. It receives ducts of a pair of large, amylase-secreting pharyngeal glands or *sugar glands*. The oesophagus leads into stomach.

Stomach. It is an irregularly shaped sturcture. The wall of stomach is lined generally by long cylindrical cells with coloured granules. It receives the two ducts of a large asymme-trically bilobed and brownish grey *digestive or liver glands*, secreting proteolytic and cellulose digesting enzymes.

Intestine. The stomach is followed by long and coiled intestine which may attain a length about fourtimes the body length. It is divided into two sections, an anterior and a posterior by a valve. The valve divides the compacted waste, enters the posterior intestine, into short faecal pellets. The valve also slows down the passage of food through the stomach and anterior intestine to allow adequate time for digestion and absorption. The posterior intestine opens out through the *anus* present in pallial groove behind the foot.

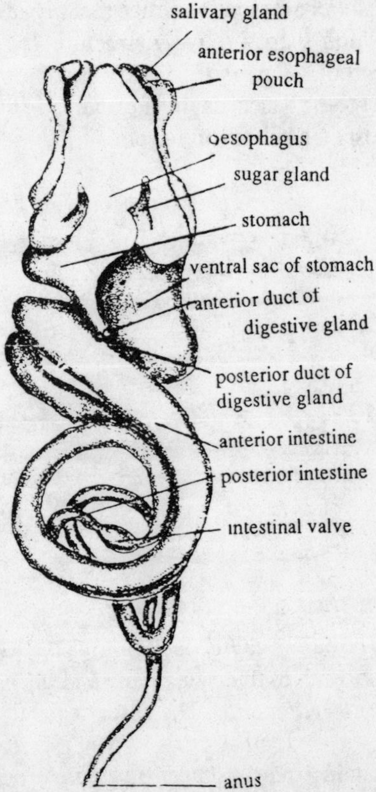

Fig. 2.4. *Chiton.* Alimentary canal.

Most chitons are microphagous and feed on fine algae and other organisms that they scrape from the surface of rocks and shell with the redula.

Digestion appears to be entirely extra-cellular. Food is stored in the form of fats, glycogen and lipids in the liver glands. Digested food is absorbed in stomach and anterior intestine.

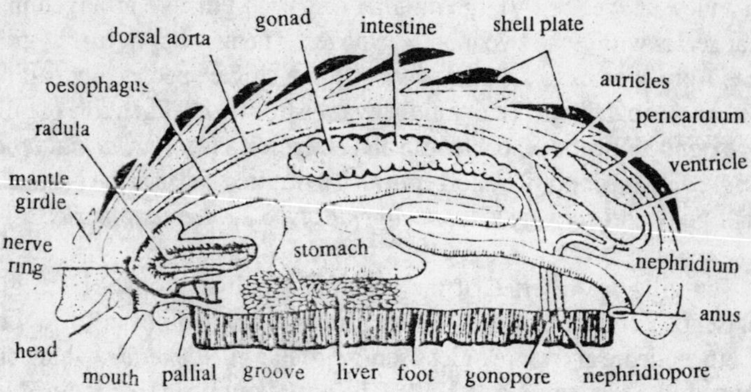

Fig. 2.5. *Chiton*. M.L.S. Showing internal structure.

Vascular System

The *heart* is enclosed in a large pericardium. It occupies the posterior dorsal region of the body. It consists of two *auricles* and a median elongated *ventricle* connected to an *aorta*. The aorta carries blood to the various organs and intervisceral blood spaces. The venous blood from the different parts of body is conducted back to a large sinus, the *afferent ctenidial channels* on either side, near the line of union of the mantle with the body. From here the blood is carried to the gills for aeration. From gills the oxygenated blood is brought to the auricle by *efferent ctenidial channels*.

The course of blood circulation may be summarized as:

→ Heart → Auricles → Ventricle.
↑ (Efferent ctenidial channels) ↓ (Aorta)

Efferent ctenidial channels ← Gills ← Afferent ctenidial channels ← Lacunae ← Organs of body ← Aorta

Respiratory System

The respiration is aquatic and takes place with the help of gills. The mantle cavity consists of a groove on each side of the body between the foot and mantle edge. A large number of paired gills are arranged in a linear series within the two mantle grooves. The number of paired gills varies from species to species even within a single species, depending upon the size of animal. Their number varies from 6-80. Each row of gills may extend along the whole length and is termed as *holobranchial* type, eg. *Boreochiton* or may be short confined to limited space at the posterior end of body and known as *mesobranchial type* eg. *Lepidopleurus*.

The gills are inserted at the bottom of pallial groove. Each gill consists of a thin central *axis* bearing an anterior and posterior row of smooth, delicate and more or less semi-circular *gill-lamellae*, which are arranged like the leaves of a book. The entire outer surface is lined by ciliated epithelium.

The gills receive the venous blood from longitudinal sinuses through the afferent ctenidial vessel. Gaseous exchange takes place on the surface of gill-lamellae. After being oxygenated in the gill-lamellae, it is carried by an efferent ctenidal vessel into the auricles.

On each side of the mantle groove towards the anterior end, the mantle margins are raised to form two inhalent openings. Water enters each inhalent chamber through these anterior openings, runs along the course of groove and passes up through the gills into each dorsal exhalent chambers. The two exhalent currents converge postesiorly and pass to the outside through one or two exhalent openings created by the locally raised mantle.

Excretory System

The excreatory organs comprises two slender *kidneys or nephridia*. Each kidney consist of a Y-shaped tube, and provided with numerous small dendritic outgrowths ramifying along its whole length among the viscera. The tube terminating blindly anteriorly. On of the paired limb of the Y, the ventral branch opens to the exterior through a *nephridiopore* into the posterior part of mantle cavity. The dorsal branch opens into pericardium through a ciliated *renopericardial aperture*.

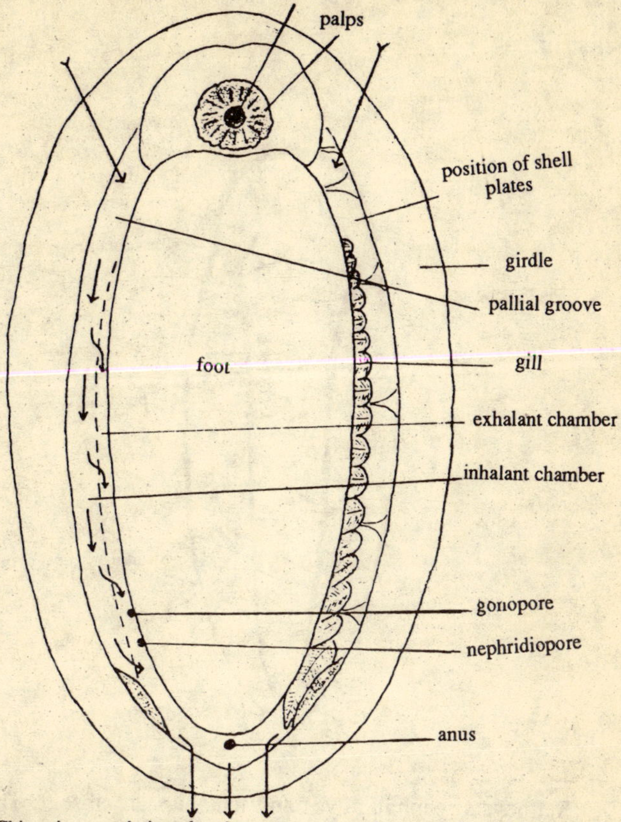

Fig. 2.6. Chiton in ventral view showing watter current by arrows in mantle groove.

Nervous System

Chitons possess the simplest and most primitive type of ladder-like nervous system which resembles the turbellarians and trematods. Ganglia are lacking or at least poorly developed. A nerve ring surrounding the oesophagus, and nerves issue form it anteriorly to innervate the buccal cavity and subredular organ. Posterisly the nerve ring gives rise to a large pair of *pedal nerve cords,* innervating muscular foot. A pair of *lateral or palliovisceral nerve cord* innervating the visceral orgnas, mantle and shell . The pedal and pallioviscesal are united with their fellow by commissures.

Sense organs

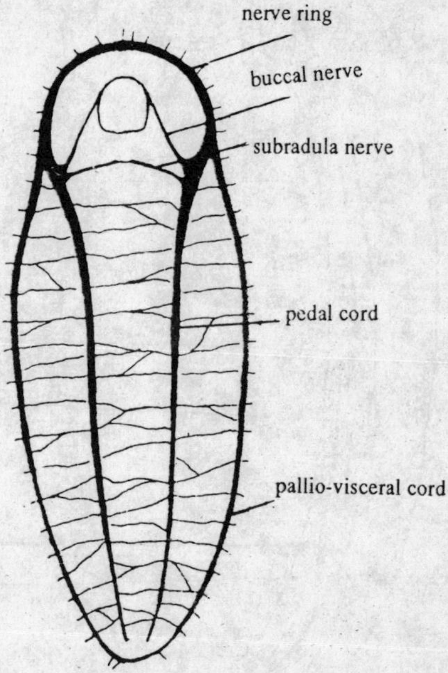

Fig. 2.7

Statocysts, cephalic eyes and tentacles are absent. There are no specialized and well defined sensory organs but according to *Haller*, typical neurosensory cells occur in the ventral epidermis of the head near the mouth. The chief sensory organs are the subradula organs and the *aesthetes*. Aesthetes, which are unique to chitons, are mantle cells that are lodged within minute vertical canals in the upper tegumentum. The canals and sensary endings terminate beneath a cap on the surface of the shell plates. The function of aesthetes is still uncertain.

In the family Chitonidae, some of the aesthetes are modified as simple eyes. These Chiton ocelli may number in thousands per individual.

Reproductive System

Chitons are unisexnal but there is no sexnal dimorphism. The gonads are paired in *Nuttalochiton* but in others the gonad is single and median.

The gonad lies dorsally above the intestine infront of pericardium. The testis and ovary are indistinguishable externally but when the products are mature, the ovary becomes greenish and the testis reddish in colour. Externally the gonad is clothed with a thin layer of muscular and connective tissue.

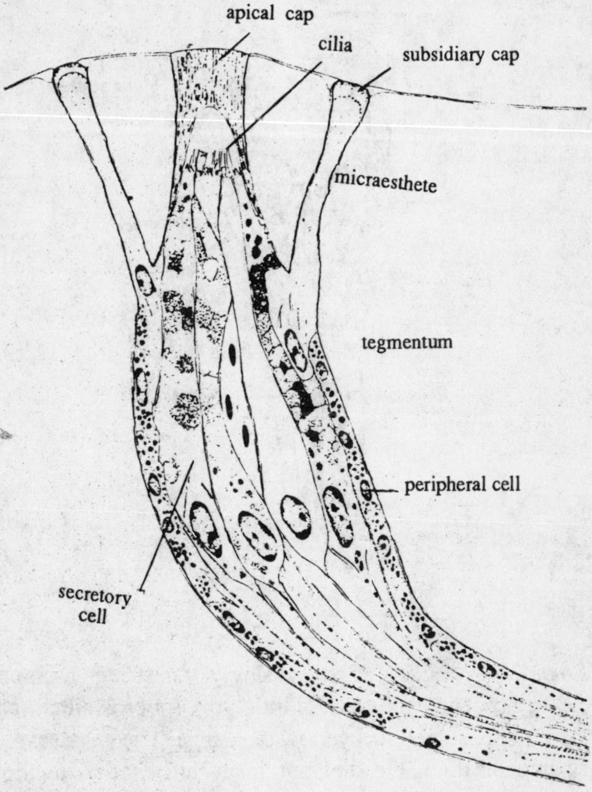

Fig. 2.8.

A pair of gonoducts arise from the dorsal surface of the psoterior end of the gonad. They proceed along the anterior wall of pericardium, then curve laterally to open into the mantle cavity infront of the renal apertures.

The copulatory organs are absent hence copulation does not take place. Sperms leave the male in the exhalent currents and fertilization occurs in the sea or within mantle cavity of female. The important factors influencing the spawning are quit water, low-tides and moonlight.

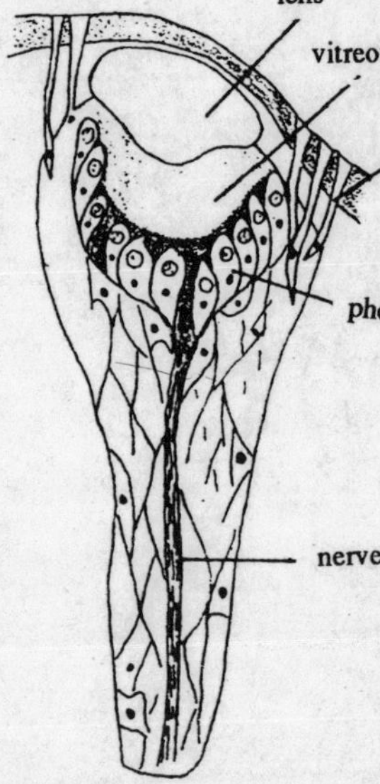

Fig. 2.9

Embryology. Eggs may be laid singly or in strings containing nearly 2,00,000 eggs. The eggs possess little yolk (microlecithal) and usually coloured green, sometimes yellow or orange. The eggs have two membranes an outer thick, the shell and inner thin vitelline membrane.

Cleavage is complete and spiral, leading to a somewhat flatenned *coeloblastula*. Gastrula, mainly of the embolic type formed by the invagination of the macromeres. The *blastopore* marks the posterior end. The gastrula soon becomes elongated in the direction of the future long axis. The blastopore gradually shifted towards the anterior end of the embryo. The ectoderm around the blastopore is invaginated to form the mouth and stomodaeum. The redular sac arises in the rudimentary form from the ventral side of the stomodaeum. By the development of an apical tuft of cilia and a double circlit of cilia or *velum*, just infront of mouth, the embryo becomes *trochophore larva*. In some cases it becomes free by cracking egg shell whereas in others it remains inside the shell upto the later stage of development.

The trochophore develops an alimentary canal consisting of a large stomodaeum ending blindly since the anus in not yet developed. The anus is formed later by invagination with the proctodaeum. Primitive *nephridia*, occuring in annelids and many molluscan trochophores, are absent. An opical platc is not present in the early trochophore but it is represented by rudiments of the cenebral ganglia which subsequently appear at the apical pole. The gills appear as papillae and gonads originate by proliferation of the anterior wall of pericardium.

At this stage, the trochophore ruptures the egg envelope and swims for about 15 minutes to several days, depending on species.

Metamorphosis. In the metamorphosis of the chiton trochophore, the post-trochal region elongates to form major past of the body. The shell gland extends up into the pretrochal region, and the shell plates are formed in the transverse grooves.The prototroch degenerates, and the animal sinks to the bottom as a young Chiton. The larval eyes are retained for some time after metamorphosis.

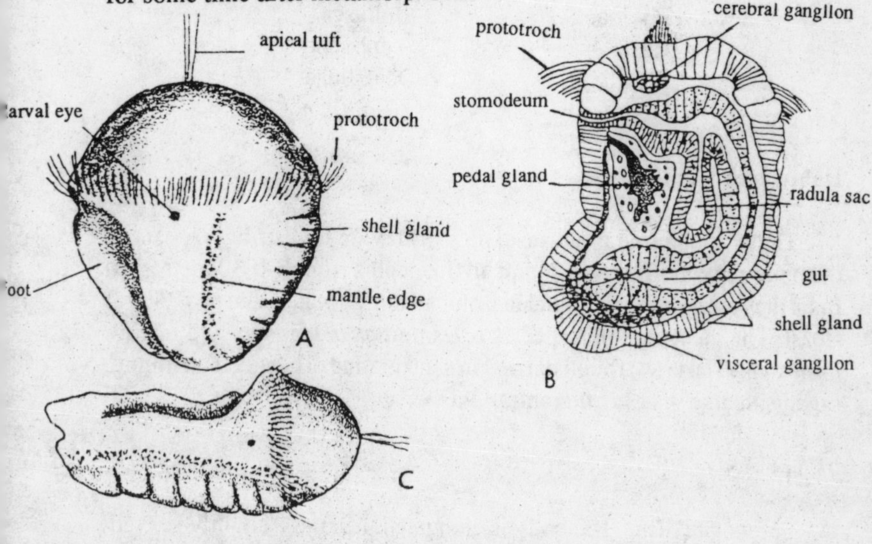

Fig 2.10

3

DENTALIUM

Dentalium is included in class Scaphopoda. Scaphopoda contains about 350 species of burrowing marine molluscs that are popularly known as 'tusk' or 'tooth shells'. These names are derived from the shape of the shell, which is an elongated cylindrical tube usually shaped like a trumpet or elephant's tusk.

SYSTEMATIC POSITION

Phylum	-	Mollusca
Class	-	Scaphopoda
Family	-	Dentalidae
Genus	-	*Dentalium*

Habits and Habitat

These are marine molluscs and world-wide in distribution except the polar regions. They are found all the depths from 6-1830 metres. In India they are commonly found in Bombay littoral waters and near Gujrat Coast. The most common species is *Dentalium octogonum* and rare *D. formosum.* These are found burried in sand or mud. They feed on minute organisms like diatoms, Foraminifera etc.

Morphology

Shape and Size. The body is elongated, headless, slightly curved, bilaterally symmetrical, concave dorsally and convex ventrally. The average length is 2-5cm. but *Cadulus mayori* does not exceed 4mm. in length. *Dentalium vernedei*, found off Japan, is the largest living species, reaching a length of 15cm.

Colouration. The shells of most scaphopods are white or yellowish, but an East Indian species of *Dentalium* is a brilliant jade green. *D. formosum* is mottled with shades of dark red.

Mantle. The body is completely enclosed by a tubular mantle, except at an anterior and a posterior apertures. The mantle cavity is large and ventral, extending continuously from one aperture to other. The posterior aperture serves for both inhalent and exhalant water currents.

Shell. The shell is external and secreted by the mantle. It is cylindrical, tubular, slightly curved open at both the ends. the shape of shell is like a trumpet or elephant's tusk. In the genus *Cadulus*, the shell is somewhat cucumber shaped. An operculum is absent and the growth takes place at the larger anterior end. The animal is attached to the shell by muscles near the posterior end.

Foot. The foot is long and cylindrical muscular organ, protrudes from the larger anterior end of the shell. It is more or less spade-like and highly extensible. The conical foot is well adapted for burrowing. It is fully extended when gorged with blood, plunges deep into the mud draws the animal by subsequent contractions. It is pointed in front, but is provided with an oblique wing-shaped fold or pleat on either side of its free extremity. In *Siphonodentalium* the foot terminates in a retractile disc with papillate margins while in *Pulsellum*, a filiform tentacle arises from the middle of the disc.

Head. Adapted to a burrowing habit, the head is reduced to a short conical projection, or proboscis, bearing the month. The eyes and tentacles are absent. In *Dentalium*, the mouth is surrounded by a rosette of palp. On the dorsal side of head, from the two lobes, arise two groups of long ciliated retractile filaments, each having a swollen, sucker-shaped tip. These are called cephalic filaments or *captacula*. These are sensory, prehensile, capable of regeneration and serve as tectile organs.

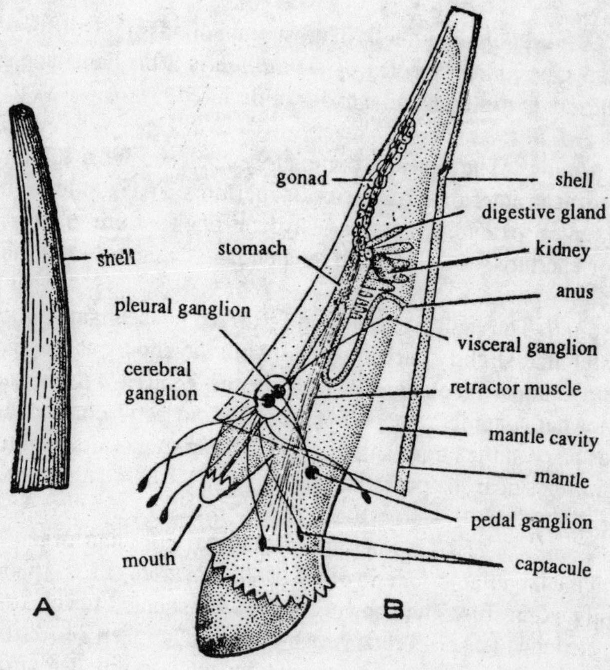

Fig. 3.1 *Dentalium*. A = shell, B = Structure

Digestive System.

The alimentary canal is somewhat straight and extends from month to anus.

Mouth. It is situated on the short oral proboscis and leads behind into the buccal cavity.

Buccal Cavity. The buccal cavity is dorsally provided with *mandible* and ventrally a *redula* and *odontophore*. The redula is short and provided with five teeth in each transverse row.

Oesophagus. The buccal cavity leads postesiorly into a short oesophagus. It is provided with two large lateral pouches. It leads into stomach.

Stomach. The stomach is simple with a vestigial caecum and gastric shield. It receives the ducts of a large bilobed symmetrical digestive gland or *liver*. The liver is situated in the upper parts of the body.

Intestine. Stomach opens into intestine. It bends forward and forms several loops and finally opens by the anus in the mid-ventral line behind the base of the foot, in the mantle cavity. An *anal gland* is present on the right side of the rectum in *Dentalium.*

Respiratory System.

The respiratory organs such as gills or ctenidia are absent. The respiration is effected by the internal surface of mantle. The mantle is thrown in the form of transverse folds. Cilia are present on ventral folds or ridges. Water enters the mantle cavity by the ciliary action through inhalant siphon and expelled out through exhalant siphon. The water is expelled after 10-12 minutes, due to a violent muscular contraction of body.

Blood Vascular System

The circulatory system is reduced to a system of blood sinuses, and there is a rudimentary heart which is represented by an enlargement of sinus near the anus. The ventricle and pericardium are absent. The blood is colourless and there is no respiratory pigment.

The main sinuses are the perianal, pedal, visceral and pallial.

Excretory system.

There is a pair of symmetrical nephridia or kidneys, situated infornt of the gonad on the ventral side of the middle region of body, one on either side of the rectum. They open into the mantle cavity by a nephridiopore on either side of the anus. They are without reno-pericardial duct and therefore do not communicate with each other.

Nervous System.

The nervous system contains the typical paired cerebral, pleural, pedal and visceral ganglia. The system is symmetrical. The *cerebral ganglia* are situated on the dorsal side of the *oesophagus* and are connected with each other by *cerebral commissure*. An *infra-oesophageal labial commissure* also arises from the cerebral ganglia.

The cerebral and pleural ganglia are connected to the pedal ganglion of the same side by a long connective which is single at its distal side but bifurcates just before reaching the cerebral and pleural ganglia sending a branch to each. Lying one on either side of the rectum, are ill defined *visceral ganglia*, united by *pleuro-viscaral connectives* to the pleural ganglia. The two pedal ganglia are situated in foot and are united with each other.

Sensory organs

There are three differentiated sensory organs:

1. *Captacula.* These are tectile and olfactory organs dorsal in position.
2. *Subredular organs.* It is a ciliated ridge on ventral side of the buccal cavity opposite to the mandible.
3. *Statocyst.* A pair of statocysts lying in the foot behind the pedal ganglia. Each contains many *otoconia.*

Reproductive System.

Scaphopods are dioecious. The unpaired gonad fills most of the posterior part of the body high up in the visceral dome. The sperms and eggs reach the outside by way of right nephridium. The eggs are shed singly and are planktonic. Fertilization is external.

Development.

The cleavage is irregular and takes place just after fertilization. The micromeres multiply much more rapidly than the large entomeres which remains for sometime undivided. The endodermic cells are invaginated to form a *gastrula* with a large blastopore at its posterior end. The gastrula is succeeded by a free swimming *trochophore* larva. It is a free swimming stage. As a result of further development it is converted into a bilaterally symmetrical *veligar* larva. The head of the larva bears a ciliated girdle or velum consisting of three rings of cilia. Its body shows a rudiment of foot ventrally. A shallow shell gland is formed on the dorsal side and extends ventrally forming two lateral mantle lobes and finally fuse togather forming the tubular *mantle* round the body.

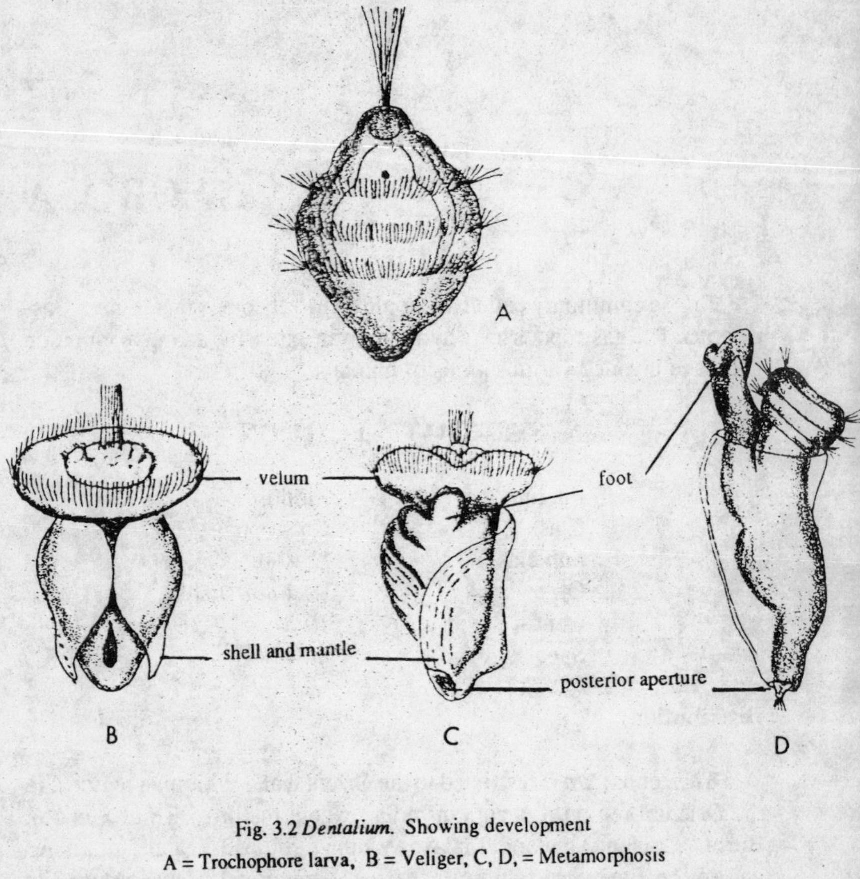

Fig. 3.2 *Dentalium.* Showing development
A = Trochophore larva, B = Veliger, C, D, = Metamorphosis

The different ganglia are formed and statocysts develop in the foot from the ectoderm. Endoderm gives rise to the stomach and intestine. As soon as the velum atrophies (5 or 6 days) the larva ceases to swim and sinks to the bottom. The mantle secretes shell.

After sometimes it begins to crawl.

4

PILA GLOBOSA

Pila is commonly called as 'apple snail'. It comes under class Gastropoda. The gastropods have undergone an extensive adaptive radiation and have invaded various kinds of habitats.

SYSTEMATIC POSITION

Phylum	-	Mollusca
Class	-	Gastropoda
Sub class	-	Prosobranchiata
Order	-	Pectinibranchiata
Genus	-	*Pila*
Species	-	*globosa*

Distribution

The genus *Pila* is restricted to the Oriental and Ethiopian regions. In the Oriental region it occurs in India (except the province of Punjab), Burma, Ceylon, Thailand, Malaya, Vietnam, Indonesia and Philippines and in the Ethiopian region, in Africa, Arabia and Madagascar. *Pila globosa* is the common species found in North India except Punjab.

Habits and habitat.

These are largest known fresh water snails. They inhibit ponds, tanks, paddy-fields and sometimes in fresh water rivers and in brackish water. They occur in those areas where there is a large amount of vegetation like *Vallisneria or Pistia*, for food. They are amphibious being adapted for life in water as well as on land. They make long journeys on land in search of areas with more abundant food and water during rains. The animal creeps by its ventral muscular foot. they pass over long periods of drought by remaining *torpid* with the aperture of shell tightly by the operculum.

External Features

Shell. The body of *Pila* is enclosed in a calcareous shell, which is somewhat conical and is coiled round a central axis, the coils, or whorls being marked off externally from each other by distinct sutures.

The top of the shell is distinguished as the *apex*. Below this are distinguishable the successive lines called *sutures* running a spiral course. The lowermost whort is considerably bigger than all the other whorls taken together, and ends in a wide opening, the mouth or aperture. In the living annimal, the aperture is directed forwards and ventrally and the apex points backwards. The internal cavity contained in the shell, is a continuous one running through all the whorls. The margin of the aperture is named the *peristome;* its outer part is called the *outer lip* and the inner part, the *inner lip* or the *columellar lip*.

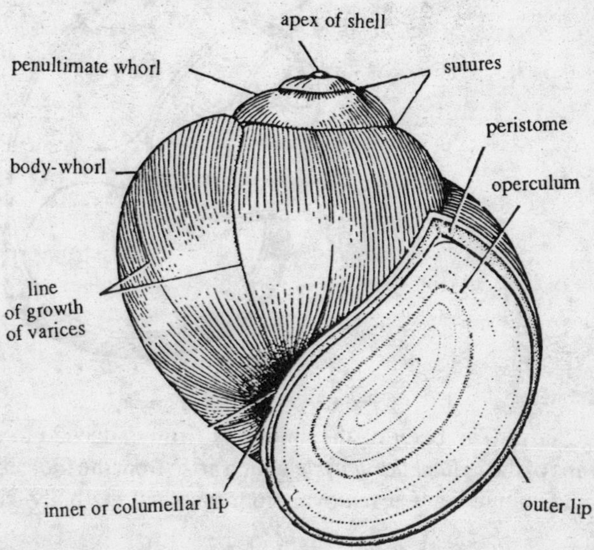

Fig. 4.1. *Pila*. Shell with operculum.

The axis of the shell or the *columella* is hollow and its cavity opens to outside by an opening called *umbilicus*, which lies close to the left corner of the inner margin of the peristome. Into the aperture fits an oblong flat plate, called the *operculum* which is composed of material secreted by certain glandular cells of the foot. The inner and outer surface of the operculum bear characteristic markings. On its inner surfaces stands out a sharp elliptic impression, called the *boss* which is produced by the attachment of the *opercular muscle*. Round the boss runs a groove. The outer surface of the operculum is marked by a number of lines of growth running round a *nucleus* which is situated subcentrally near the inner margin.

Microscopic structure of the shell. The shell of *Pila* consists of three layers. The outer layer is known as *periostracum*. It is greenish brown. The underlying two layers are made up of *calcareous plates*. In the *ostracum*, the middle layer, the calcareous plates run at right angles to the margin of the aperture of the shell. *Ostracum* is the thickest of the three layers. In *hypostracum*, the innermost layer, the plates are arranged parallel to the margin of aperture. The structure of the two inner layers is similar, but the calcareous plates in the two zones lie at right angles to one another.

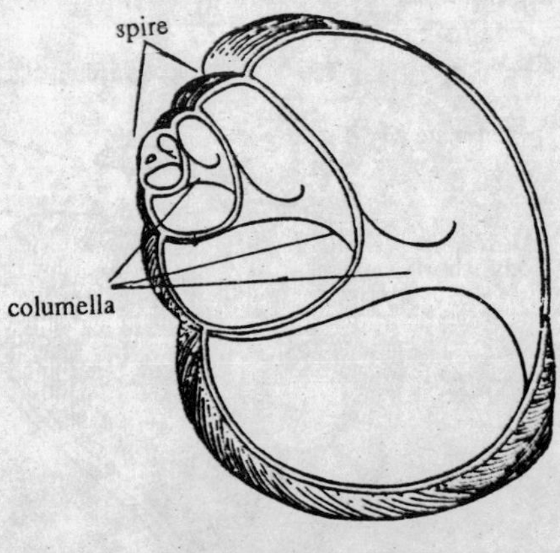

Fig. 4.2. *Pila*. Shell in section

Body. The body of the animal is firmly attached to the shell by means of the *columellar muscle* which arise from the foot and in inserted on the columella. On removal from the shell the body can be distinguished into foot, head and visceral mass. The head and foot are bilaterally symmetrical while the visceral mass is asymmetrical.

Head. The anteriormost part of the body, overhanging the foot, forms a definite head. It bears dorsally a pair of small tapering and contractile *first pair of tentacles or labial palps.* The snout bears the mouth in its anterior surface below the palps. Behind the labial palps are the longer, filament like *true tentaclesor second pair of tentacles.* These are hollow and very contractile and become about 5cm. long when fully extended, but only 1.5cm. when contracted. The eyes are placed on tips of short stalks or *ommatophore* which arise from the outer side of the tentacles near their bases. On the two sides of head connecting it and mantle, occur two fleshy structures, which are called *nuchal lobes* or *pseudepipodia.* The left nuchal lobe is better developed and forms the *respiratory siphon.*

Foot. The large, strongly muscular, ventral part of the body forms the *foot.* When fully expanded, it is roughly triangular in shape, with the apex directed backwards. The foot has a broad, flat, smooth and grey ventral surface, or the *sole,* for creeping. The foot bears the *operculum* dorsally on its posterior end, called the *operculiferous lobe.* When the foot is withdrawn, the operculum completely fits into the mouth of the shell and closes it.

The head-foot complex is attached to the visceral mass by a short inconspicuous *neck.*

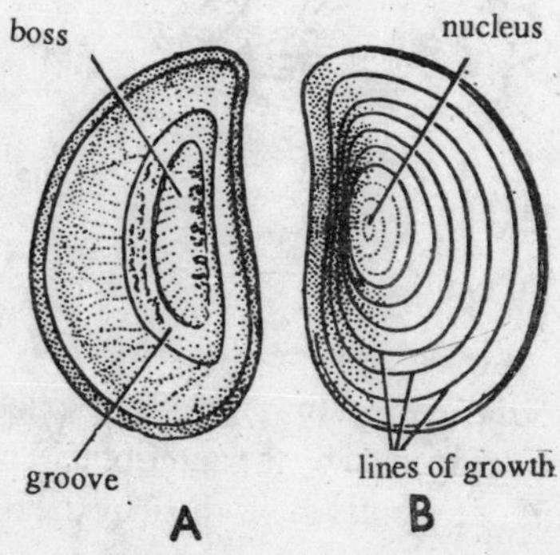

Fig. 4.3. *Pila.* Operculum A = inner view B = Outer view

Visceral mass. The visceral mass is a spirally coiled hump-like structure on the dorsal side and contains all the vital organs of the body.

Mantle. the skin of the *visceral mass* form the *mantle* or the *pallium*, the characteristic molluscan structure. The free edge of the mantle is thickened and possesses the shell-secreting glands. The mantle keeps a regular check over the anterior part of animal. There is a long, narrow *supramarginal groove* along the thick free edge of the mantle. Behind the *supramarginal groove,* there is a slight, thickened band, which has the shell-secreting glands.

The shell is secreted by the *mantle*. The *periostracum* and the *ostracum* are secreted by the *supra-marginal groove* and the shell-gland respectively. The *hypostracum* is secreted by the general epithelial covering of the mantle, which also serves as an additional organ of respiration.

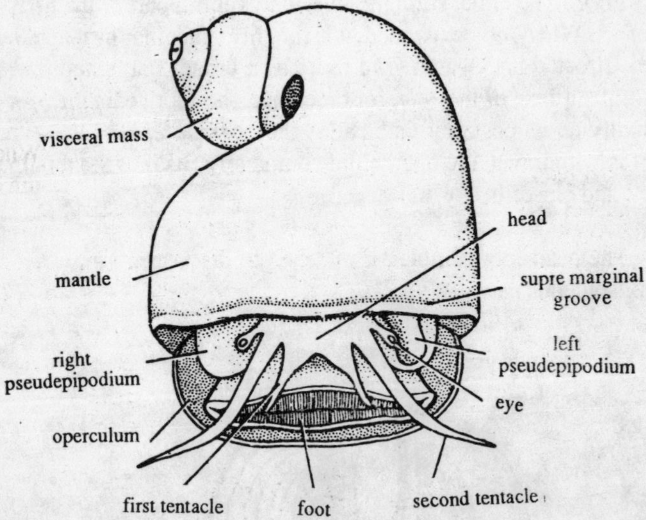

Fig. 4.4. *Pila*. Body after removing shell.

Mantle Cavity or Pallial Cavity. This is the large space whose roof and greater part of the sides are formed by the anterior part of mantle and the floor and remaining part of the sides are formed by the dorsal surface of the body behind the head. In this cavity a number of organs are present known as *organs of pallial complex.* The organs of pallial complex are:

1. *Epitaenia*-When a longitudinal incision is given along the left of the ctenidium a prominent ridge, the *epitaenia,* is seen on the floor of the mantle cavity. It arises near the right nuchal tube and extends upto the extreme posterior end of the cavity. It divides the mantle cavity into a small right *branchial chamber* and large left *pulmonary chamber*.

2. *Ctenidium or gill*-It lies at the extreme right side of the mantle cavity hanging vertically downwards from its dorso-lateral wall. Being *monopectinate* (come-like), its gill filaments or *lamellae*, which are somewhat triangular in shape, hang freely in the branchial chamber.

3. *Rectum*-It lies on the left of ctenidium on the floor of mantle cavity and opens to the exterior by *anus,* situated near the right *pseudepipodium* or right *nuchal lobe*.

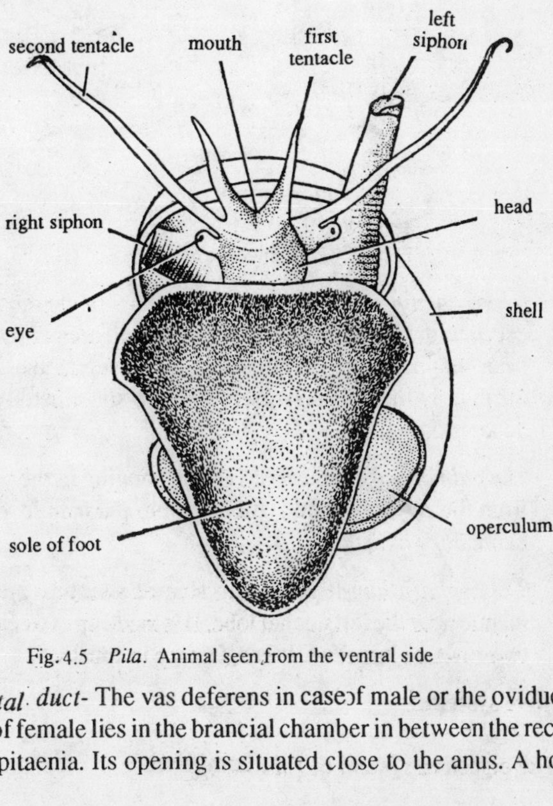

Fig. 4.5 *Pila*. Animal seen from the ventral side

4. *Genital duct*- The vas deferens in caseɔf male or the oviduct in case of female lies in the brancial chamber in between the rectum and epitaenia. Its opening is situated close to the anus. A hook-

like *penis* can be seen in front of the genital pore in case of male specimen. At the base of penis is found the *hypobranchial gland.*

5. *Hypobranchial gland-* It is a glandular thickening at the base of the penis. Its secretion helps in the act of copulation.

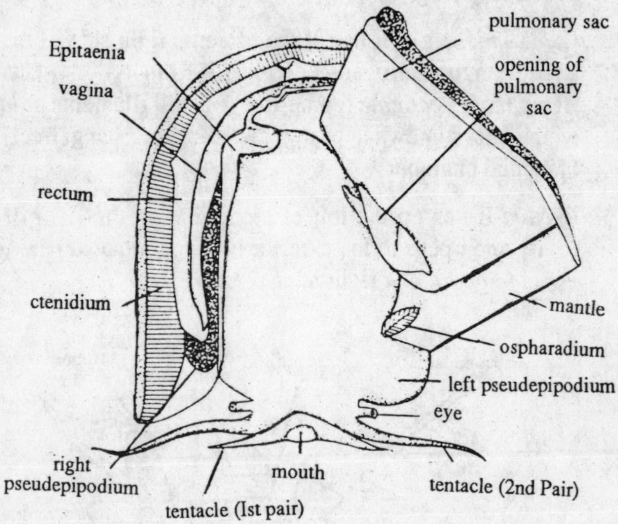

Fig. 4.6. *Pila*. Structure of mantle cavity.

6. *Renal opening.* The anterior chamber of the renal organ projects into the branchial chamber as a small area of reddish colour, near the posterior termination of the epitaenia. Its external opening is like a transversely oblique slit situated in a shallow depression.

7. *The pulmonary chamber-*It is a sac hanging in the mantle cavity from the roof of mantle. It opens into the mantle cavity by the *pulmonary aperture.*

8. *The ospharidium-*It is a comb shaped structure attached to the mantle near the left nuchal lobe. It is made up of a central axis and two rows of lamellae. It is *olfactory* in function.

Digestive system.

The digestive system of *Pila* comprises:

I. Alimentary canal

II. Digestive glands.

I. *Alimentary canal.* The alimentary canal is tubular and peculiar due to the twisting of visceral mass. It is divisible into three parts, fore-gut, mid-gut and hind-gut.

1. *Fore gut.* It is also called stomodacum and is ectodermal in origin. It is made up of buccal mass and the oesophagus.

Buccal mass. Mouth is a median vertical slit situated at the tip of snout and is surrounded by frilled folds of integument. True lips are absent. These frills serve as secondary lips. Mouth leads into the buccal cavity, which is the cavity of buccal mass. It is greatly obliterated and contains a pair of jaws and radula. The jaws divide the buccal cavity into

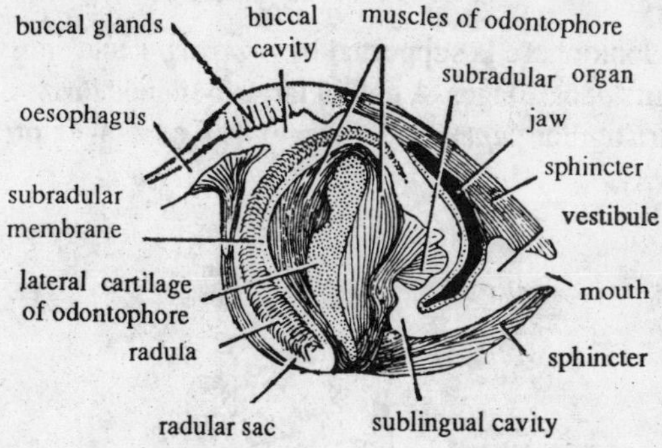

Fig. 4.7 *Pila*. V.L.S. of buccal mass.

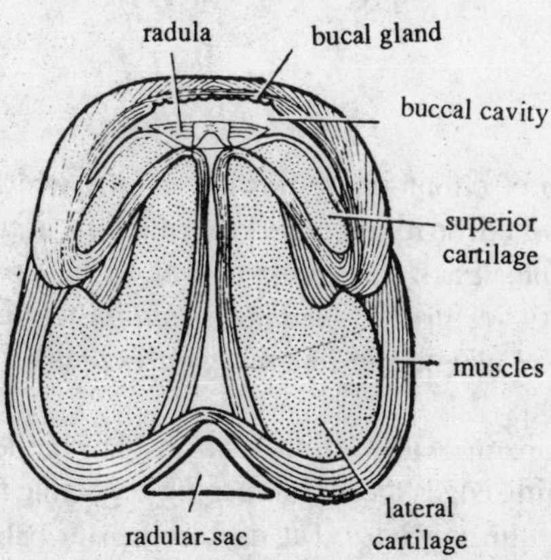

Fig. 4.8. *Pila* Buccal mass in T.S.

two sections, the *anterior* also called *vestibule* and a *posterior*. The anterior chamber is extremely short. Its posterior limit is marked by the presence of a pair of bands, which are especially thickened areas of the internal cuticular lining of the buccal cavity. In the posterior region of buccal cavity there lies an *odontophore* or *tongue mass* on the floor of the buccal cavity. On the roof of buccal cavity occur the dorsal buccal glands.

Fig. 4.9. *Pila*. Radula

The odontophore is supported by *protractor* and *retractor muscles* and two pairs of cartilages. A pair of large S-shaped *lartal* cartilages and a pair of triangular *superior cartilages.*The cartilages project into the buccal cavity.

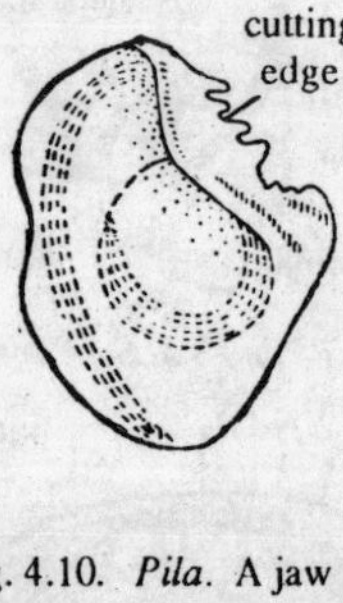

Fig. 4.10. *Pila*. A jaw

At the top of odontophore lies the radula, which is a star-shaped body and bears teeth on its upper surface which are arranged in transverse rows as in a file. Each transverse row of radula contains seven teeth including a large *central* or *median rachidian* tooth, followed by one small lateral and two needle-like *marginals* on either side.

The teeth are made up of chitin reinforced by hardened protein. They bear sharp cutting edges meant for rasping vegetable food. The old teeth are replaced by the new ones. The new teeth grow behind the radula and push it forward.

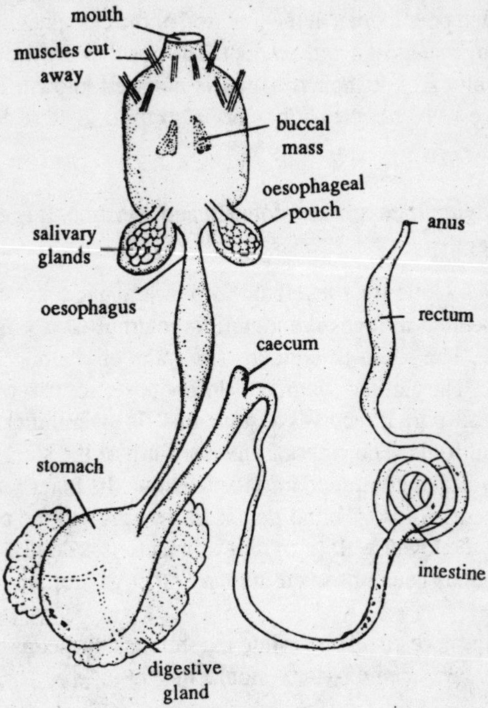

Fig. 4.11 *Pila*. *Alimentary canal*.

The number of radular teeth in a row and the size and shape of teeth vary in different genera and species.

The two chitinous jaws form the posterior limit of the *vestibule*. The jaws lying dorsolaterally in the region of the sphincter muscle of the buccal mass are connected together by a cuticular membrane. The anterior cutting edge of each is *truncate* and *serrate* and bears two or three large teeth. Their movements are brought about by the sphincter muscles of the buccal mass. They appear to be formed by simple local thickenings of the cuticular lining of these areas.

The buccal cavity leads into oesophegus.

Oesophegus. It is like a long and narrowtube. It starts from the dorsal part of buccal mass and extends backwards along the median line to open into stomach after entering the visceral mass. A pair of round cream-coloured *oesophegeal pouches* are found in the lateral sides of oesophegus. Each pouch open at the junction of the oesophegus and the buccal cavity by means of a narrow duct. The pouches are the projections of oesophageal wall but their function is not well known. Probably they secrete digestive enzymes. They are also used to store food for some times.

2. *Midgut.* It comprises stomach and intestine. It is endodermal in origin.

Stomach. It lies on the left side of visceral mass, below the pericardium, embedded in digestive gland. Its internal cavity is U-shaped in appearence. The stomach contains two parts or chambers: the *cardiac* and *pyloric*. The cardiac chamber is left or posterior part of stomach and forms its main part. It receives oesophagus. Its inner lining is thrown into longitudinal folds. The right or anterior limb of the stomach is *pyloric chamber* which is continued into the intestine. Its inner lining is thrown into transverse folds. A blind pouch-like outgrowth the *caecum*, originates from the lower wall of pyloric stomach. It is devoid of crystalline cone which has been observed in many other gastropods.

The digestive gland opens into the stomach by means of a duct at the junction of cardiac and pyloric stomach.

Intestine. The pyloric stomach followed by a narrow coiled tube, the *intestine*. In the beginning the intestine runs backward, then curves upward and takes 2 1/2 or 3 coils between the gonad infront and the digestive gland behind, before joining the rectum.

3. *Hindgut.* The hind-gut or proctodaeum is ectodermal in origin, comprising rectum.

Rectum. It forms the terminal part of alimentary canal. It is a thick walled tube running into the branchial chamber and opens to exterior by the *anus* lying about 0.6 mm. from the edge of the right nuchal lobe.

II *Digestive glands.* A number of digestive glands are associated with alimentary canal. These are:

Salivary glands. Salivary glands are paired, white branching glands lying one on either side of the oesophagus behind the buccal mass. From each salivary gland a duct enters the *buccal mass* to open into it in the region of the dorsal buccal glands.

They secrete mucus and enzyme. Mucus helps in the lubrication of the radula and the smooth transport of food. The enzyme helps in digestion of starch.

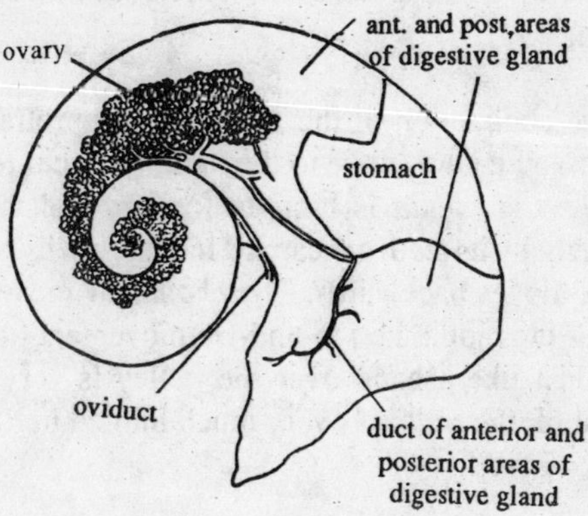

Fig. 4.12. *Pila*. Inner view of digestive gland.

2. *Digestive gland.* The digestive gland is large spirally coiled somewhat brownish or dirty-green in colour. It forms the greater part of the visceral mass. It consists of two lobes, the smaller is connected with the stomach, the bigger extends to apex of the spiral of the visceral mass. From each lobe arises a main duct, the two ducts unite to form a single duct before opening into the stomach. These ducts, into the digestive gland, branch repeatedly and end blindly in a very large number of small tubes, the *alveoli*. The alveoli are lined with a digestive epitheliummade up of three types of cells-(i) *secretory cells* which secrete a cellulose digesting enzyme, (ii) *resorptive cells* which digest proteins intracellularly, and (iii) *lime cells* or calciferous cells which store calcium phosphate. Semi-digested food enters into these alveoli, where digestion of cellulose and proteins takes place.

This structure is regarded as '*liver* by most of the zoologists but as it performs the functions of the various digestive glands and also is the main organ for the absorption of the digested food, therefore, *Dakin* has given the term *digestive gland* to this structure.

3. *Buccal glands.* A pair of elongated glands is present in the roof of buccal cavity. They secrete digestive enzymes.

4. *Oesophegeal pouches.* As already described, they probably secrete digestive enzymes.

Food and feeding. The food consists chiefly of aquatic succulent vegetation. It is also reported that sometimes *Pila* feeds on dead animal tissues too.

By the contraction of the sphincter and protractor muscles of the buccal mass, the jaws move to the mouth and cut thin pieces of plants. Meanwhile, the *radula* is brought forward and the pieces of aquatic plants caught by its teeth are carried into the cavity of the buccal mass as the radula moves backwards. The *chain saw* movement (deseribed by *Huxley*) of the radula i.e., to-and-fro movement of the radula over the cartilage just like a band over the pulley is very characteristic. The movement of the *radula* is very much limited in *Pila* in comparison to other *gastropods.*

Digestion

The salivery glands pour their secretion into the buccal cavity where it mixes with the food. It converts insoluble starch into soluble sugar. In the stomach the food is digested by the secretion of the digestive gland. The digestive juice brings about the extracellular digestion of food, whereas intracellular digestion takes place inside the *resorptive cells* of digestive gland. The resorptive cells digest the cellulose portion of the food. Absorption of digested food takes place in the digestive gland and the intestine. The undigested food passes out from the rectum through the anus. From anus it comes into the branchial chamber and from there it comes to the exterior along with the out-going current of water

Respiratory System.

Pila is an amphibious animal so that it employes both aquatic and aerial methods of respiration.

Organs of respiration. The organs of respiration consist of a pair of leafly expansions of mantle situated in the pallial cavity and called *ctenidia, pulmonary sac* and a pair of *nuchal lobes.*

1. *Ctenidia*. It consists of numerous thin *lamellae* attached to the mantle wall by their broad bases, but have their apices free in the branchial chamber. The line of attachment is knownn as *ctenidial axis*. A *ctenidium* with such an axis with a single row of lemellae is described as *monopectinate gill*. The lemellae present in the middle of the ctenidium are largest. They gradually become smaller towards the ends. Each lamella has a longer left side known as *efferent side* being on the side of efferent ctenidia sinus and the relatively smaller right or *afferent side* lying on the side of afferent ctenidial sinus.

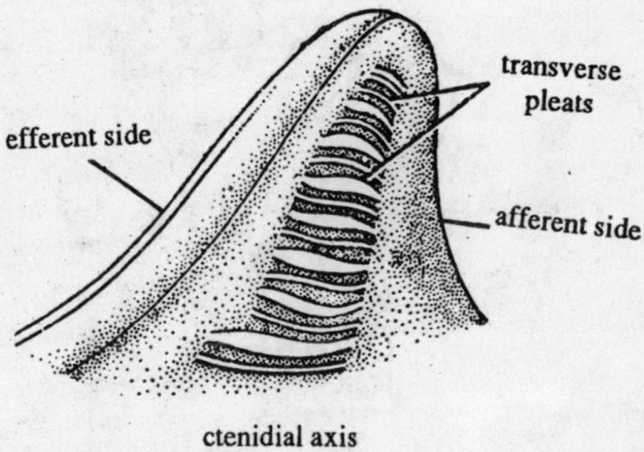

Fig. 4.13. *Pila* A gill lamella.

Shallow transverse ridges, called *pleats* are found in the anterior and posterior regions of each lamella. Each lamella arises as a double-layer of epithelium internally supported by connective tissue and muscle fibres. The epithelial layer consists of three kinds of cells:

(i) Non-ciliated columnar cells

(ii) Ciliated columnar cells

(iii) Glandular cells.

Underlying the epithelium there is a thin basement membrane below which lies connective tissue layer consisting of cells with a few scattered nuclei and large number of obliquely running muscle fibres. After the muscle fibres is situated a layer of pallial epithelium.

The ctenidium of *Pila* is situated on the right side of the body but the distribution of nerves in it and the position of ospharidium shows that

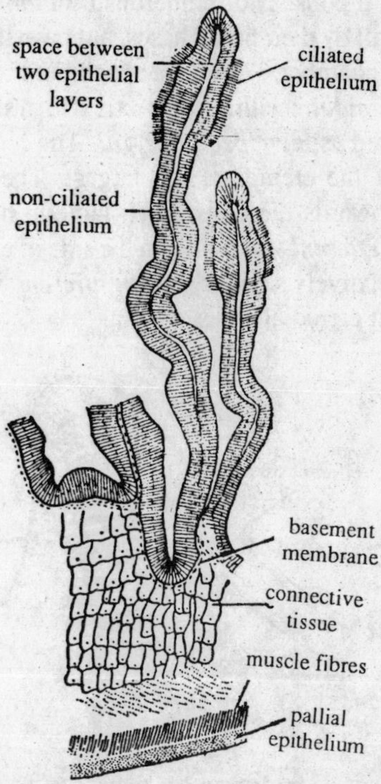

Fig. 4.14. *Pila*. T.S. of two gill lamella.

gills of the left side has been pushed to the right side by extensive development of the pulmonary sac.

2. *Pulmonary Sac.* It is also known as *lung*. It is a sac-like structure hanging from the roof of mantle cavity and is entirely closed except for a large opening with which it opens into the mantle cavity. It is developed from the mantle itself and consists of a densely pigmented dorsal wall and a creamy white ventral wall. The flaps on the two sides of the opening are of unequal size, the left being larger than the right and help in controlling the opening of pulmonary sac. The outer wall of the pulmonary sac consists of an external epithelial layer containing large numbers of fine black pigment granules. This is followed by a basement membrane below which lies a thick layer of connective tissue and finally the endothelium of the inner surface.

According to some workers the pulmonary sac is modified second ctenidium but this has not been proved to be correct.

3. *Nuchal lobes.* The mantle is provided with two fleshy and contractile *nuchal lobes* or pseudopipodium on the two sides of head over the foot. The left nuchal lobe is well developed whereas the right is less developed. They form two respiratory tubular funnels longer and smaller during respiration.

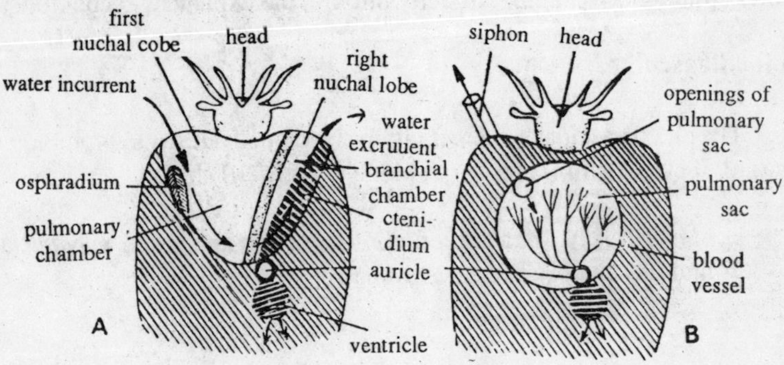

Fig. 4.15. *Pila*. Mechanism of respiration

A= Water current in aquatic respiration. B = Air current in aerial respiration.

Mechanism of respiration.

Aquatic respiration. Aquatic respiration usually takes place, when the animal lies submerged in the water, attached to the plants or floating in muddy water. During this process, the animal is fully expanded, and the two nuchal lobes, which are equally expanded, appear more or less like shallow channels. The *left nuchal lobe* after a time increases in size and takes the form of a *gutter*. The respiratory current of water enters through the gutter-like *left nuchal lobe* flowing beneath the *osphradium* and reaches the posterior part of the *pulmonary chamber* of the mantle cavity and then crossing the *epitaenia* it enters the branchial chamber, where it bathes the entire length of the gill allowing exchange of gases and, finally, passes outside through the *right nuchal lobe.*

Aerial or pulmonary respiration-The aerial respiration occurs either when the pond water is poor in O_2 or *Pila* is making excursion on land.

Pila moves towards the surface of water and thrusts out its left nuchal lobe which becomes elongated and tubular to form the respiratory siphon. The pulmonary chamber becomes greatly enlarged and is completely cut off from the branchial chamber. The air enters the pulmonary sac through respiratory siphon and the respiratory movements produced by the alternate contraction and expansion of the pulmonary sac help in inspiration.

While *Pila* is on land, the respiratory siphon is not formed and the air enters pulmonary chamber directly through the expanded nuchal lobe.

Blood vascular System

The blood vascular system is highly developed due to its amphibious nature. It consists of (I) heart (II) blood vessels (III) blood.

I. *Heart*. It is two chambered consists of an *auricle* and a *ventricle* and lies enclosed in *pericardium*.

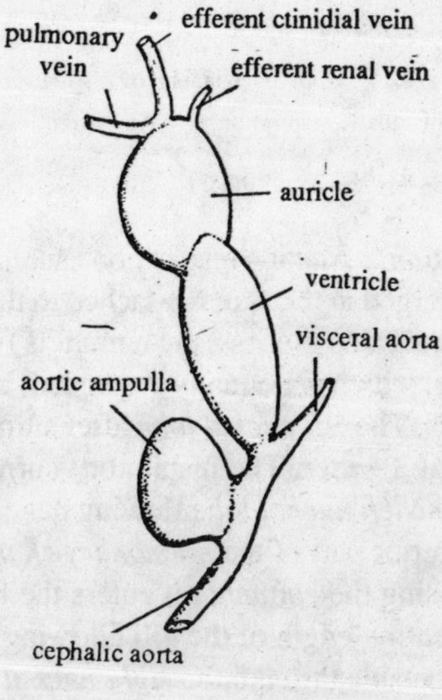

Fig. 4.16. *Pila*. Heart

(i) *Pericardium*. The pericardium is thin walled, oval chamber enclosing the heart and situated between the pulmonary chamber and the posterior renal sac on the left side of the body whorl. Posteriorly, it is surrounded by the stomach and the digestive gland. The renal sacs are present on its dorsal side while the oesophagus is found on its ventral side. The pericardial cavity is the true coelom as it communicates with renal organs and the reproductive organ develops from its wall in early stages of development. Pericardium communicates with the posterior renal chamber by *reno-pericardial aperture.*

(ii) *Auricle*. The auricle is a thin walled, roughly triangular sac highly contractile in nature. It is situated on the dorsal side of pericardium. It receives blood through three main veins-(a) *efferent ctenidial vein* (b) *efferent renal vein* and (c) *pulmonary vein*. At the lower end it is connected with ventricle through the *auriculo-ventricular aperture*. The aperture is guarded by two semilunar valves which allow the blood to flow from auricle to ventricle but not in reverse direction.

(iii) *Ventricle*. It is an ovoid sac lying below the auricle. It has thick, spongy and muscular walls and a reduced cavity due to a coarse meshwork of muscular strands. From the lower end of the ventricle arises a large artery, the *aortic trunk*. The opening between the ventricle and aorta is guarded by two semi-lunar valves which prevent the flow of blood back into the ventricle.

II. *Blood Vessels*. It includes arteries, sinuses and veins.

1. *Arteries*. The aortic trunk divides immediately into two branches an anterior *cephalic aorta* and a posterior *visceral aorta*.

(a) *Cephalic aorta*. There is a thick-walled *ampula* at the base of the cephalic aorta. It is highly elastic and helps in pumping the blood. The Cephelic aorta, beyond the ampulla, gives off immediately three arteries on the outer side.

(i) A *cutaneous artery* which supplies the skin.

(ii) A pair of *oesophegeal arteries* which supply the oesophagus.

(iii) A thick *pallial artery* which supplies the left side of mantle, left nuchal lobe and osphradium.

(iv) A *pericardial artery* arises from the inner side of the cephalic aorta. It supplies blood to the pericardium.

The main aortic trunk after giving off above branches, reaches the porterior renal chamber and supplies both the renal chambers by many branches, the *renal arteries.*

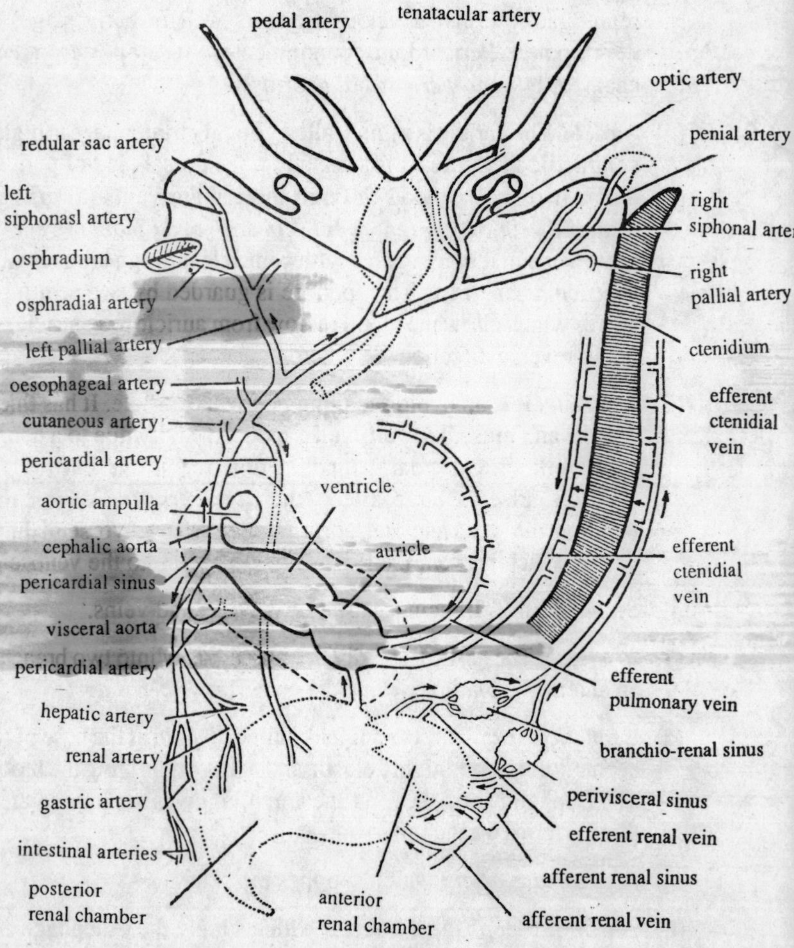

Fig. 4.17. *Pila*. Heart and blood vessels

The main branch of the cephelic aorta now runs along the left side of the oesophegus for a short distance and then crosses over to its right. Here it appears as a large artery and gives off a number of small branches to the oesophagus and the floor of mantle cavity. On its right side it divides into three branches.

(i) *Right pallial artery* to the right side of the mantle.

(ii) *The right siphonal artery* to the right nuchal lobe.

(iii) *The penial artery* to the copulatory organs.

The main cephalic aorta again divides into following four branches:

(i) A *redular sac artery* to the redular sac.

(ii) A pair of *optic nerves* to the eyes.

(iii) A pair of *tentacular arteries* to the tentacles.

(iv) *Pedal arteries* which branches to form a network in foot.

(b) *Visceral aorta*. It passees through the visceral mass and supplies blood to the visceral organs by the following branches:

(i) *A pericardial branch*, which supplies the pericardium, the anterior part of the digestive gland and the skin.

(ii) A large *gastric branch*, which supplies the different parts of the stomach.

(iii) A number of small *intestinal arteries*, which supply the intestine.

(iv) A large *digestive branch*, which supplies the digestive gland and the gonad.

(v) Many *renal arteries* to the anterior and posterior renal chambers.

In the end the visceral aorta terminates into branches which supply the rectum and anus.

2. *Sinuses*. The ultimate branches of the arteries open into wide spaces without walls, called the *blood sinuses*. Some of them act as blood capillaries and contact the arteries with the veins. Other sinuses join together to form larger sinuses. In *Pila* there are four sinuses:

(i) *Perivisceral sinus*-It occupies the anterior part of the body above the foot and below the floor of mantle cavity. It surrounds the anterior part of alimentary canal.

(ii) *Peri-intestinal sinus*-The peri-intestinal sinus is confined around the coils of the intestine.

(iii) *Branchio-renal sinus*-It lies along the right side of the anterior renal chamber.

(iv) *Pulmonary sinus*-It is confined to the walls of the pulmonary sac.

3. *Veins*. The veins carry the venous blood from the various parts of the body to the auricle directly or through the respiratory and excretory organs.

(i) *Afferent ctenidial vein*-It collects blood from the perivisceral sinus, branchio-renal sinus, rectum and the terminal part of the genital duct and carries it to the ctenidium, where it branches to supply blood to the branchail lamellae for purification.

(ii) *Efferent ctenidial vein*-It collects oxygenated blood from the lamellae of ctenidium, mantle and copulatory organ and carries to the auricle.

(iii) *Afferent renal vein*-It receives blood from the peri-intestinal sinus and empties into the posterior renal chamber.

(iv) *Efferent renal vein*-It orginates from the posterior renal chamber and empties into the auricle.

(v) *The pulmonary vein*-It collects blood from the pulmonary sac and carries it to the auricle.

III. *Blood*. The blood is colourless and contains colourless amoebocytes in the plasma. the blood of *Pila* has a light bluish tinge. A respiratory pigment, the *haemocyanin* is found dissolved in the plasma. The haemocyanin helps in respiration.

Course of Circulation.

The ventricle pumps the mixed blood through branches of cephalic aorta and visceral aorta to various parts of the body. The venous blood is collected from the various parts into the perivisceral and periintestinal sinuses. From these sinuses the blood is sent back to the auricle either directly through kidneys for the removal of nitrogenous waste products or indirectly to the gill and mantle during aquatic respiration. The auricle thus gets aerated and non-areated blood from these organs and it is the mixed blood which is forced into the ventricle to be pumped again to various organs of the body.

Excretory System.

In *Pila globosa*, there is a single large *renal organ or kidney or organ of Bojanus* lying behind the pericardium. The excretion is done by the left-kidney as the right kidney has modified into the gonoduct. On one side it opens into the pericardial cavity and on the other to the outside through the mantle cavity.

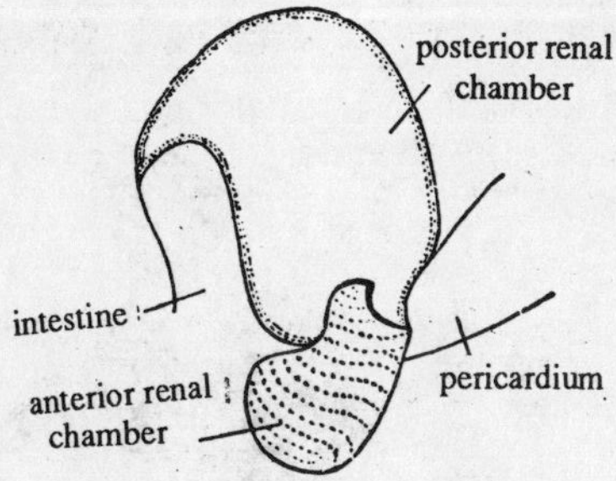

Fig. 4.18. *Pila*. Renal Chamber

The renal organ is made up of two distinct parts: a right *anterior renal chamber* and a left *posterior renal chamber*.

1. *Anterior renal chamber*. It is oval, reddish organ, lying on the dorsal side of the animal in front of the *pericardium* and *posterior renal chamber*. The cavity of the anterior renal chamber is very small due to the presence of a number of triangular leaf-like processes or *lamellae* are arranged on either side along a thick median longitudianl axis, or the *efferent renal sinus*. On the floor of the chamber, the *lamellae* are arranged on either side of a median axis, the *afferent renal sinus*. The anterior renal chamber communicates with the posterior renal chamber through a small opening at one end and opens by an elongated opening into the mantle cavity lying to the right of the *epitaenia*, at the other end.

2. *Posterior renal chamber*. It is hook like broad chamber brown or grey in colour situated behind the anterior renal chamber, in between the rectum on the right and the pericardium and the digestive gland on the left. Its large internal cavity encloses a part

of the genital duct and a few coils of the intestine. At one end, it communicates with the anterior renal chamber through an aperture, and at the other with the pericardium through an elongated slit-like *reno-pericardial aperture*, perforating a thin vertical *reno-pericardial septum*, separating the two. The afferent and efferent renal vessels profusely branch in the roof of this chamber.

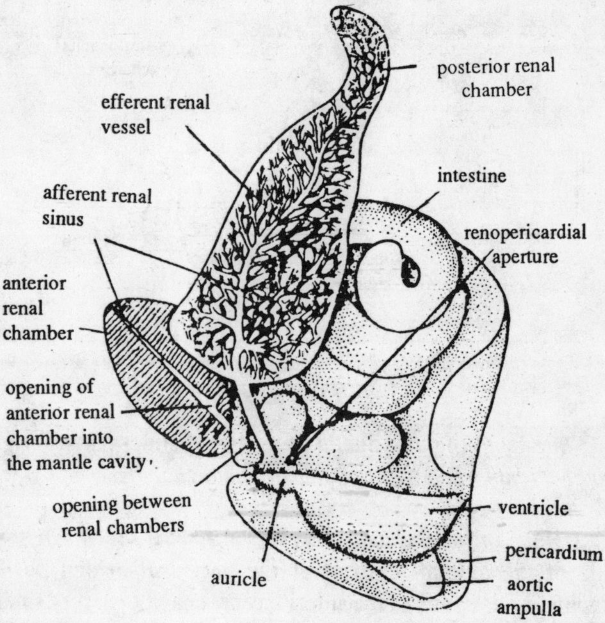

Fig. 4.19. *Pila*. Excretory System.

Physiology of excretion. The renal organ separates the nitrogenous waste from the blood. The excretory fluid passes from the posterior renal chamber to the anterior renal chamber from where it is discharged into the mantle cavity through the external renal aperture and finally it goes out with the outgoing current of water through right siphon. The excretory matter contains chiefly ammonia compounds, urea and uric acid. For couserving water, *Pila* converts ammonia into insoluble uric acid.

In most of the gastropods, the digestive gland also forms an organ of excretion. It contains some excretory cells which take up excretory waste products. These are eleminated by way of stomach and intestine.

Nervous System.

A well-developed nervous system. It possesses two characteristics (i) except the visceral ganglia all other ganglia form a ring-like structure around the buccal mass (ii) due to torsion the visceral loop is twisted to form the shape of '8'. The nervous system of *Pila* includes the following ganglia, their commissures and connectives.

1. One pair of cerebral ganglia.
2. One pair of buccal ganglia.
3. Paired pleuro-pedal ganglionic mass.
4. One pair of visceral ganglia
5. Unpaired supra-intestinal ganglion.

1. Cerebral ganglia. A pair of large, flat and conical cerebral ganglia are situated on the dorso-lateral sides of buccal mass. These are connected together by thick, band-like *cerebral commissure* dorsally. In addition to this both these ganglia are also connected with each other by thin *labial commissure* lying below the cerebral commissuse.

Each cerebral ganglion is connected with the *buccal ganglia* of its side by *cerebro-buccal connective* and by *pedal* and *pleural* ganglia by a thick band-like *cerebropedal* and *cerebro-pleural connectives*. All these connectives are situated on the lateral sides of buccal mass.

Anteriorly the cerebral ganglia give nerves to the skin of snout, tentacles and buccal mass. Posteriosly, they give nerves to tentacles, eyes and the statocyst.

2. *Buccal ganglia.* They are two in number. They are situated on the dorso-lateral side of the buccal mass at its junction with the oesophagus. The two buccal ganglia are connected with each other by a stout buccal commissusre, which runs on the ventral side of the oesophagus behind the buccal mass. Each ganglion is

also connected with the cerebral ganglion of its side by the cerebro-buccal connective.

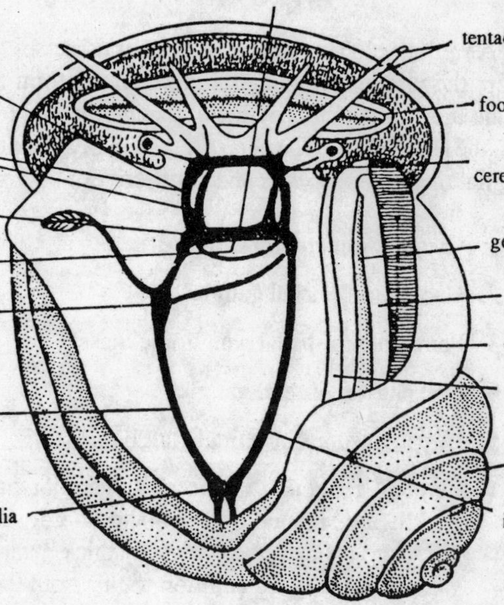

Fig. 4.20. *Pila*. Nervous System

The nerves are supplied from each buccal ganglion to the posterior part of *buccal mass* and *radular sac*.

3. *Pleuro-pedal ganglia.* A pair of large, somewhat triangular ganglionic masses present one on either ventro-lateral side of the buccal mass. Each one is formed by the fusion of an outer *pleural* and an inner *pedal* ganglion, separated by a faint notch. The right pleuropedal ganglionic mass also consists of the *infra-intestinal* ganglion fused with it.

The two pedal ganglia are connected with each other by two *pedal commissures* which are placed one above the other. These commissures are situated below the buccal mass. Many nerves are given off to the foot from the anterior and posterior parts of the depal ganglion. Behind the pedal commissures, the two pleural ganglia are connected with each other by *infra intestinal nerve*. The left pleural ganglion gives off nerves to the mantle, ospheridium, left nuchal lobe, columellar muscles and the anterior part of gill. The right pleural ganglion supplies nerves to parietal wall, epitaenia, right nuchal lobe, copulatory organs, rectum, genital duct and columellar muscles.

4. *Supra-intestinal ganglion*-It is a fusiform ganglion situated on the left side 1/4th of an inch away form the pleuro-pedal mass of its side. It is connected anteriorly to the left pleural ganglion, posteriorly to the left visceral ganglion by thick nerves and on its inner side to the *sub-intestinal ganglion* of right side by a thin nerve. The thin nerve is known as *supra-intestinal nerve*. On its outer side the supra-intestinal nerve issues a short but thick nerve to the mantle and the anterior part of the ctenidium.
5. *The visceral ganglion.* It is formed by the fusion of two spindle-shaped ganglia. It lies at the base of the visceral mass. It is connected with the *right pleural ganglion* by a thin nerve, the *infra-intestinal visceral connective* and with *supra-intestinal ganglion* by the left supra-intestinal visceral connective. It supplies nerves to the renal organ, genital organ and to the stomach and digestive gland.

Sense Organs

The entire exposed surface of *Pila* is provided with sensory cells but followings are special sensory organs:

1. *Ospheridium.* A small ospheridium is present in aquatic gastropods while absent in terrestrial snails. It was specially described by *Spengel* (1881), *Bernard* (1890), *Demal* (1954). It is unpaired in *Pila* and situated dorsoventrally on the roof of mantle cavity. It is bipectinate structure hanging like a curtain in the course of the respiratory water current. It consists of a specialized and usually elevated and ciliated region of the epithelium in which these is an accumulation of the sensory cells. The leaves of ospheridium are arranged on both the sides of a *central axis*. Each leaf is just like a double fold of the wall of central axis.

In a transverse section, the osphradium consists of an outermost covering of a single layered *epithelium*, internally lined by a thin *basement membrane,* the interior filled up with *nerves, connective tissue and blood spaces.* The epithelial cells are elongated possessing basal nuclei, and they are of three types-(i) *sensory,* (ii) *ciliated* and (iii) *glandular.* The ciliated cells line the attached margin, while the sensory cells devoid of cilia cover the osphradium. the flask-shaped glandular cells are found scattered among the sensory cells.

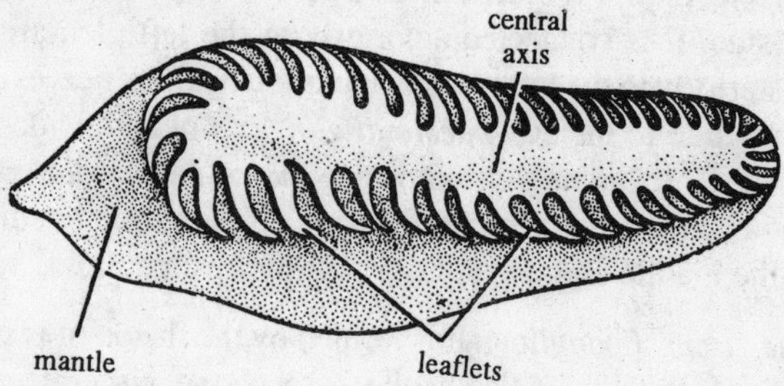

Fig. 4.21. *Pila.* Ospheridium

Formerly it was regarded as a rudimentary gill, but it has now been established that the ospheridium serves to test the chemical and physical properties of water. The ospheridium helps in the selection of food material (*Spengel*).

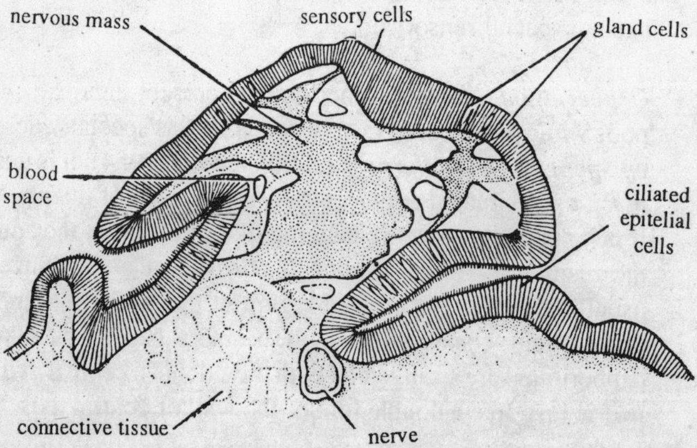

Fig. 4..22 *Pila.* T.S. of Ospheridium.

2. *Eyes.* Two eyes are borne short stalks called *ommatophores* situated behind the second pair of tentacles.

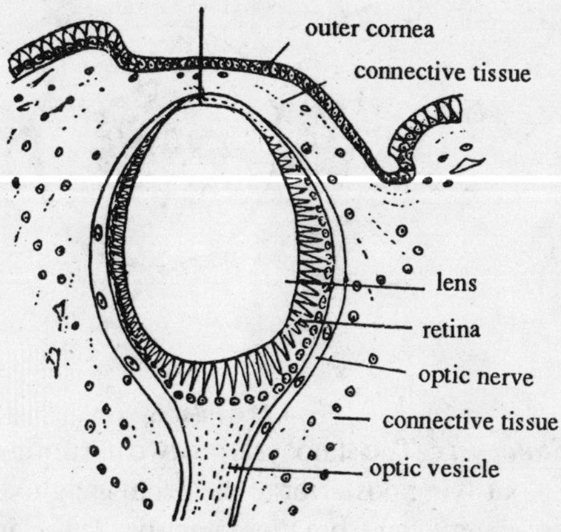

Fig. 4.23. *Pila*. H.L.S. of eye.

In section the eye is in the form of a pyriform cup or *optic vesicle*. The outer epithelium of the eye-stalk continues over the eye forming the *cornea or pellucida externa.* It consists of a row of squarish cells devoid of pigments. Lying below the cornea is a thick layer of connective tissue made up of fibres and nuclei. This layer encloses the pyriform *optic vesicles* consisting of a transparent epithelium forming *inner cornea or pellucida interna* lying next to connective tissue. The ovoidal structureless *lens* occupies the central space of the vesicle. The sensitive part of the eye of the *retina* forms the outer coat of the vesicle made up of large cells filled with pigment granules. Retina consists of large pigmented cells, which are of two types: (1) large, broad cells with hair-like processes on their outer ends, the *visual cells* and (2) the compress cells lying in between the visual cells, the *supporting or packing cells.* Both kinds of cells at the inner ends are connected with nerve-cells and branches of optic nerve, which enter the optic vesicle at its posterior end. Internally, the retina is bounded by a basement membrane.

The eyes of *Pila* do not play an important role for sight. The snail seems to be unable to distinguish objects but only responds to the changes in the intensity of light and can detect quick movements.

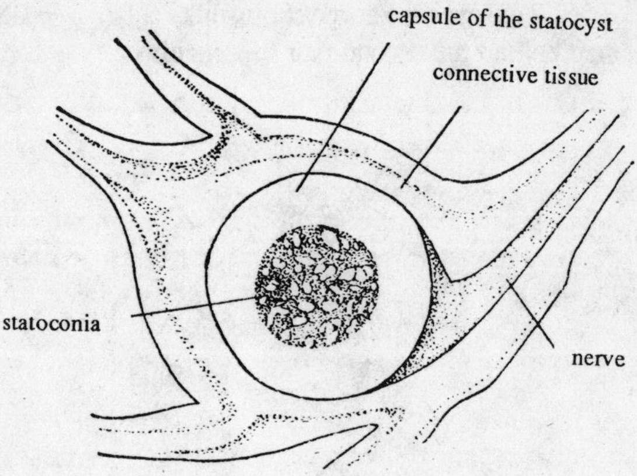

Fig. 4.24. *Pila*. Statocyst

3. *Statocysts*. The statocysts are two pyriform sacs of creamy white colour lying posterior to the pedal ganglion of each side and are not innervated from these ganglia. These are connected to cerebral ganglia by a pair of slender statocyst nerves. The statocyst is filled with fluid and contains a single large *statolith* or few to many smaller bodies called *statoconia* floating in the fluid. Formarly it was supposed to be the organ of hearing and hence it was called *otocyst* but it is proved that these are the organs of equilibrium *(Delage)*

4. *Tentacles*. There are two pairs of cephalic tentacles in *Pila*. The second pair of tentacles are *rhinophore or olfactory organs*.

Reproductive System

Pila is unisexual and a well marked sexual dimorphism is seen. The shell of male is smaller than that of female. In male penis is well developed and the female has a more swollen body whorl than the male.

Male reproductive System. The male reproductive organs comprises:

1. Testis
2. Vasa efferentia
3. Vas deferens

4. Seminal vesicle
5. Penis
6. Hypo branchial gland.

1. *Testis.* It is a single, flat, plate-like and roughly triangular whitish structure, occupying the upper part of the first two or three whorls of the shell. It lies in close association with the brownish or greenish digestive gland along its upper and inner or columellar sides and appears quite distinct on account of its creamy colour. A thin cutaneous membrane separates the testis from the shell.

The germinal epithelium of testis produces two types of speerms.

(i) *Eupyrene sperms.* These are small and thread-like. Their body consists of a head and a tail. These are motile and fertilize the ora. They measure 25μ long.

(ii) *Oligopyrene sperms.* They are large semicircular or spindle-shaped. The head is large and the tail bears 4 or 5 cilia. These are non-motile and incapable of fertilizing ova.

2. *Vasa efferentia.* From the testis arise several delicate ducts called vasa efferentia. They unite towards the posterior edge of testis to form a common duct, the *vas deferens.*

3. *Vas deferens.* The vas deferens commences from the posterior part of the testis and passes under the skin and is differentiated into two parts:-

(i) *Proximal tubular part*-It forms posterior part of testis under the skin along inner side of the digestive gland upto the posterior renal chamber and finally turns towards the left to approach the pericardium. Here it opens into the terminal glandular part of the vas deferenss.

(ii) *Terminal glandular part*-It is the distal thick-walled glandular part of the vas deferens which runs along the left side of the rectum into the branchial chamber and opens into it by male genital pore. The male genital pore is situated a little behind the anus to a claw-like projection of *genital papilla.*

4. *Seminal vesicle.* It is situated on the right of the pericardium at the junction of the anterior and posterior renal chamber. It is

swollen flask-shaped structure whose distal end is closed. The semenal vesicle stores the sperms and opens into the vas deferens at the junction of its tubular and glandulalr parts.

5. *Penis or Copulatory organs.* On the left side of the inner surface of the mantle, there is a thick glandular flap of yellowish colour. It is attached along its right side but its left free edge is rolled to form spoutlike penis-sheath. Inside the sheath is the penis.

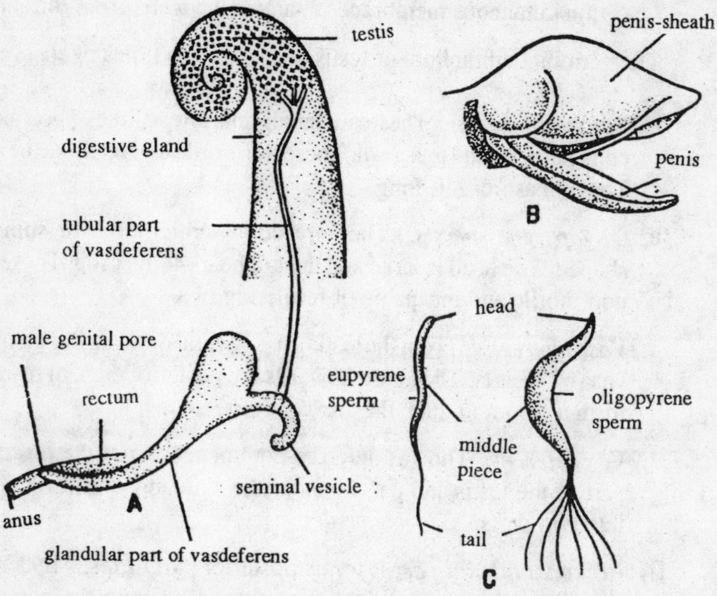

Fig. 4.25. *Pila.* A = Male reproductive organs, B = Copulatory organ C = Sperms

The penis is about 1.25 cm. long, provided with flagella arising from the roof of the mantle. It is slightly covered with swollen base and tapering end has a groove along its length on the inner side. It is capable of great extension.

6. *Hypobranchial gland.* It is a glandular structure with pleated surface at the base of penis-sheath. The gland is without a duct and, therefore, its seeretions are poured directly upon its surface.

Female Reproductive Organs. The female reproductive organs are following:

1. Ovary
2. Oviduct
3. Seminal recepticle
4. Uterus
5. Vagina
6. Copulatory organ
7. Hypobranchial gland.

1. *Ovary.* The ovary is present in the same position as the testis in the male. It is found in two or three whorls of the shell. It is orange coloured structure remains attached to the inner surface of the digestive gland. Branches of the ovary are single layered, flask-shaped structures with slender tubular necks uniting together to form minute ducts which open into oviduct.

2. *Oviduct.* The oviduct is a semi-transparent, narrow tube arising from the middle of the ovary and passing downwards along the margin of the digestive gland. Finally, it ascends upwards to receive the opening of receptaculum seminis.

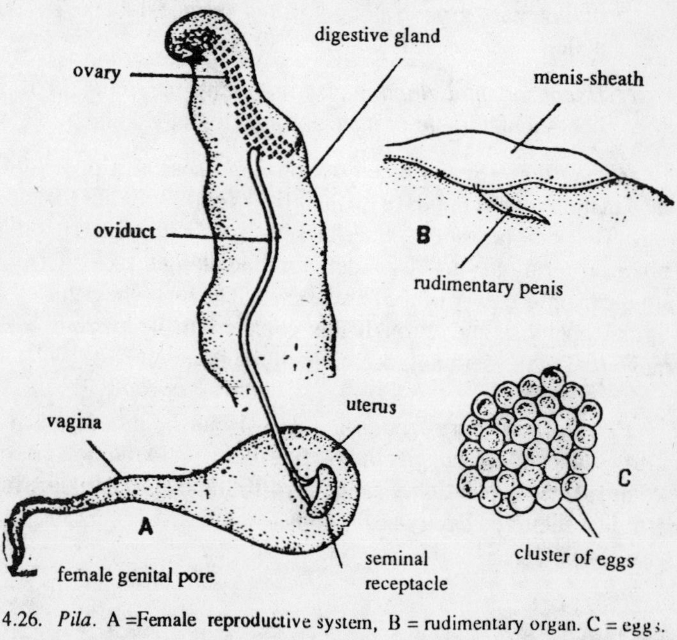

Fig. 4.26. *Pila.* A =Female reproductive system, B = rudimentary organ. C = eggs.

3. *Seminal recepticle.* The small, beanshaped *receptaculum seminis* or seminal receptacle lies enclosed within the cavity of the posterior renal chamber, closely attached ventrally to a thin-walled pouch from the walls of the uterus, called the *pouch of the receptaculum.* It is meant for storing the sperms received from the male *Pila* during copulation.

4. *Uterus.* It is a large, yellow and pear-shaped structure situated below the intestine in the body whorl. It is differentialed into a broad rounded *basal part* and a narrow tubular *apical part.* Seminal recepticle opens into the basal part while apical part receives the vagina.

5. *Vagina.* The terminal narrow part of the oviduct is known as vagina. It is a cream-coloured tube which runs forward on the right side of rectum to open into the branchial chamber by *female genital aperture* situated behind the anus.

6. *Copulatory Organ.* The rudimentary penis is not enclosed within a well-developed *penis-sheath,* but lies beneath a glandular fold of the mantle. It is a thin, flageller structure nearly 6mm. long and uniform in thickness but pointed at the tip. It also bears a rudimentary groove along its inner surface, like the penis of the male.

7. *Hypobranchial gland.* It also represents the rudiment of the hypobranchial gland of male as an incipient glandular thickening.

Copulation. Snails breed during rainy season. Copulation takes place in the water of ponds or on their banks and continues for 3 hours or more. The penis becomes connected at the base with genital papilla at the end of the terminal part of vas deferens and along the sheath is inserted into the mantle cavity of the female snail. The tip of the penis enters the vaginal opening, and through this connection the spermatic fluid is transferred to the seminal recepticle of the female.

Fertilization. Fertilization is internal but the development of the embryo takes place outside the body of female. Ovulation takes place two days after copulation. Female lays 200-280 eggs at a time in moist earth in a sheltered cavity near ponds and lakes.

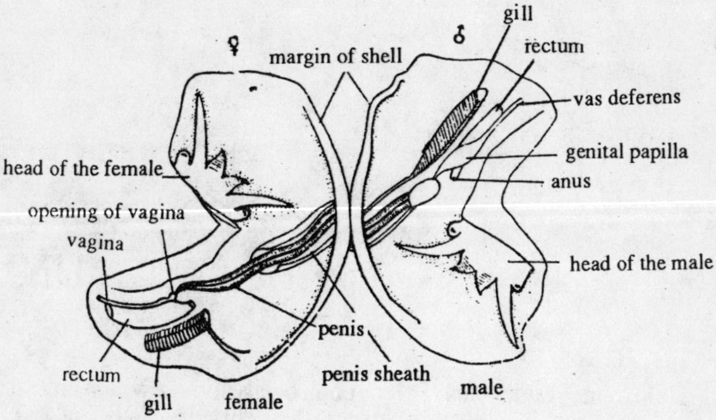

Fig. 4.27. *Pila*. Showing Copulation.

Development. The eggs are rounded and as big as the pea-seeds. Each egg consists of an outer whitish shell, enclosing a double shell membrane, a thick mass of solid white albumen and a little central mass of liquid albumen containing the embryo. During development the visceral mass and the shell of embryo become spirally coiled and a characteristic phenomenon of *torsion* takes place resulting in the asymmetry of the body. The young snails emerging from the eggs are similar to the adults.

5

UNIO

Unio comes under class Pelecypoda or Bivalvia in which an external and internal symmetry is seen unlike gastropods. The cephalic region is rudimentary and the body is laterally compressed.

SYSTEMATIC POSITION

Phylum	-	Mollusca
Class	-	Pelecypoda
Order	-	Eulamellibranchiata
Genus	-	*Unio or Anodonta*

Distribution

The 'freshwater mussel' is a familiar representative of phylum Mollusca. The family Unionidate to which *Unio* belongs is cosmopolitan in distribution. The family comprises several genera and about 1000 species of which a good number are represented in India. *Anodonta* is the common representative of Britain, while in India *Unio* and *Lamellidents marginalis* are of common occurance.

Habits and habitats

It is found in rivers, streams, lakes and ponds. During day time it usually lives partly burried in mud or sand at the bottom, keeping its posterior end exposed, It comes out to shallow places in night. They may also occur wedged in betwwen the rocks and stones, with the valves slightly spread and the two siphons exposed. It slowly creeps with a ploughshare-like muscular foot. It hardly covers 16 kilometres in a year. Sexes are separate but there is no sexual dimorphism externally. Development is indirect through a larval stage and takes place in brood chamber, which are enlarged water tubes of female's gills.

External features

Shape and size. It is laterally compressed. It is elongated and enclosed and protected in a bivalve shell. The size varies from 2" to 4" in length.

Shell

A typical bivalve shell consists of two similar more or less oval. Usually convex valves which are attached and articulate dorsally with each other. Each valve bears a dorsal protuberance called the *umbo* which rises above the line of articulation. The umbo is the oldest part of the shell, and the concentric lines around it are lines of shell growth. The two valves are attached by an elastic protein band called the *hinge ligament*, which is covered above with the periotracum. The hinge ligament is so constructed that when the valves are closed, the dorsal part is stratched and the ventral or inner part is compressed. Thus, when adductor muscles relax, the ligament causes the valves to open. The inner

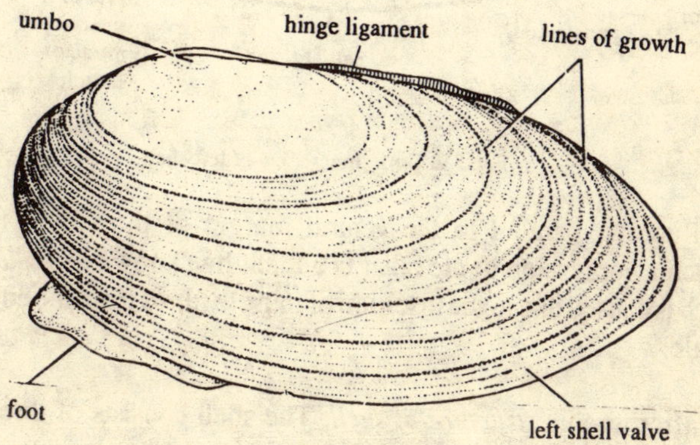

Fig. 5.1 . *Unio* External feature

surface of each valve possesses dorsally along the hinge line, small sharp ridges and teeth-like projections separated by groove or sockets, These are called *hinge teeth*. The teeth of one valve fit into the corresponding sockets of the other to prevent lateral slipping. However, a dorsal fold of mantle tissue still lies between the tooth and socket of two opposing valves.

The valves of the shell are pulled together by two large dorsal muscles, called adductors. The posterior muscle being larger than that of the anterior. Near the impression of the anterior adductor muscles are two small impressions, the dorsal and posterior one left by *anterior retractor muscle* and a ventral and posterior by protractor muscle. A small impression of a *posterior retractor muscle* also lies dorsal to the impression of posterior adductor muscle.

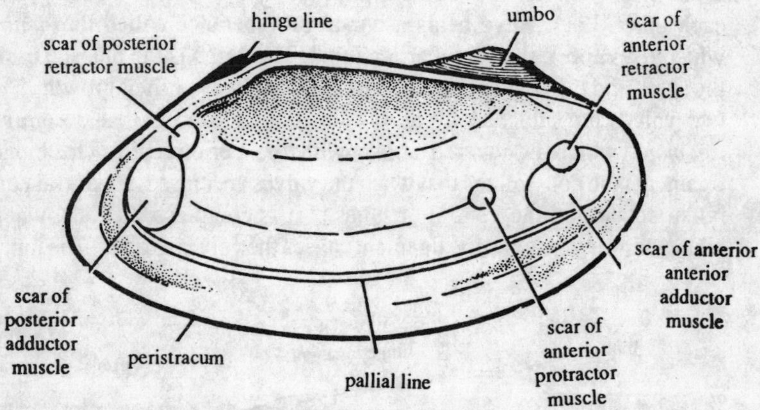

Fig. 5.2. *Unis*. Inner view of left shell.

Running parallel to the ventral margin of the valves from one adductor impression to the other is a thick line known as *pallalial line*, which is caused due to the insertion of muscle fibres from the edges of the mantle.

Microscopic structure of shell: The shell consists of three distinct layers:

1. *Priostracum*. It is the outermost thin trnslucent and greenish

brown layer formed of a special chitinous substance, the *conchiolin*. The colour of the shell is due to this layer. It is secreted by the edge of the mantle and is a protective layer for the underlying calcareous prismatic layer which otherwise can be eroded out by the action of carbolic acid of the surrounding medium. The periostracum is worn away from the older part of the shell, like umbo, leaving the prismatic layer exposed to the exterior.

2. *Prismatic layer*. It is the middle layer and is crystelline. It is made up of calcium carbonate separated by thin layers of conchiolin. It is arranged perpendicularly to the surface of shell. It is also secreted from the edge of mantle.

3. *Nacreous layer or pearly layer*. It is the innermost layer of shell also known as the *mother of pearl or nacre,* since it presents a smooth, iridescent surface. It is secreted by the whole outer suface of mantle and is in the form of alternative transverse layers of $CaCo_3$ and conchiolin. The layer of both the types run parallel to the shell surface.

Formation of shell.

The shell is made up of calcium carbonate (89-99%), very little of phosphate of lime (1-2%) and organic base called the conchiolin. Two theories have been proposed regarding the formation of shell.

1. *Bowerbank and Carpenter's Theory*. According to *Bowerbank* and Carpentor, the shell is formed by the growth of interstitial deposits like that of formation of teeth and bones in higher vertebrates.

2. *Reaumur - Eisioz Theory.* According to this theory the shell is formed by the cuticle-like deposition on the outside of skin . It is formed by minute calcareous particles which bounds together. Mantle takes important part in the deposition of shell. Every portion of mantle bears shell deposition cells. These cells separate the carbonate of lime from blood, which is the main constituent of shell. The umbo portion of shell is secreted first and subsequent layers are added in gradual succession forming the lines of grwoth.

The edge of the mantle bears three folds an inner, middle and outer. The outer fold is related to secretion of the shell. The inner surface of outer fold lays down the periostracum and the outer surface secretes the first calecareous layer. The entire mantle surface secretes the remaining caleareous cells.

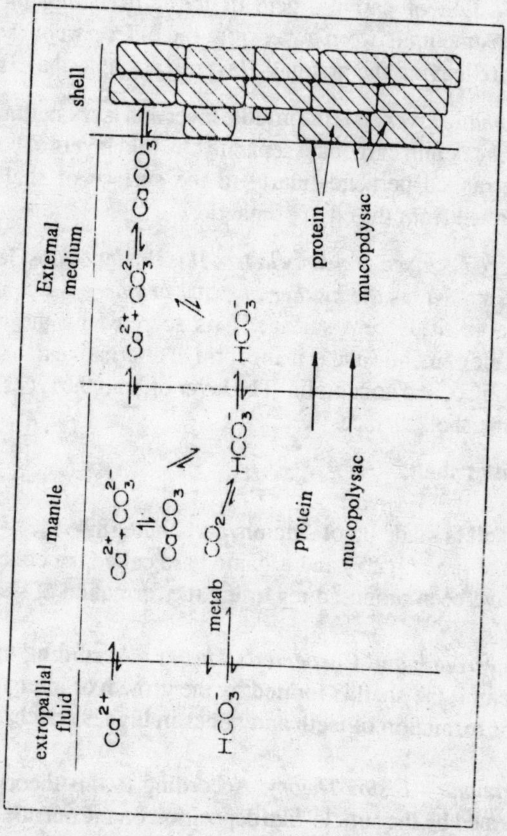

Fig. 5.3. *Unio.* Process of shell secretion.

Mantle

When the shell is removed, the soft body of the animal is exposed, covered in a thin, semi-transparent soft fold of skin, called mantle or pallium. It consists ot two lateral halves or mantle lobes or folds, which correspond to the two valves of the shell. These lobes are thin sheets of tissues closely attached to the inner suface of the valve. The lobes come off from the dorsal part of the body-wall, while, ventrally and anteriorly, they are free, posteriorly, they are fused in the middle to from a *dorsal* and a *ventral siphon.* These siphons allow water in and out. Currents of water enter through the ventral or incurrent or inhalant siphon. The ventral siphon is formed by approximation of mantle lobes and its edge are provided with small tentacles. The water leaves through the dorsal or *excurrent or exhalant siphon.* On the posterio dorsal side, the two mantle lobes also form a *dorsal - mantle pore histologically, the mantle consists of :*

1. An outer layer of columnar epithelium beset with numerous unicellular glands secreting nacre.

2. A middle fibrous connective tissue.

3. An inner ciliated epithelium containing mucous secreting cells.

Mantle cavity or Pallial cavity.

The space enclosed between the two lobes of mantle is called mantle cavity. It encloses *viseral mass, gills, foot* etc. which are exposed when one fold is removed.

(i) *Visceral mass.* It is occupies the dorsal part of the mantle cacity. It is dark in colour and contains various organs such as digestive system, circulatory system., excretory system and reproductive system. A greenish-brown *digestive gland* is present in the anterio-dorsal region. A pericardial cavity containing the heart in the mid-dorsal region and a dark coloured paired kidneys below the pericardiun.

(ii) *Gills.* Gills are a pair of long, thin, double plate like structures, which hang freely into the mantle cacity on each side from the visceral mass. There is sieve-like arrangement in the gills with minute apertures, covered by cilia. The line of attachment of the gills to the visceral mass forms a horizontal partition dividing the mantle cavity in a large ventral infra-branchial chamber and a small dorsal suprabranchial or cloacal chamber.

(iii) *Foot.* The foot is a muscular structure projecting from the ventral surface. Its size and form are very variable depending upon the life of the animal. The mass of the foot is invaded by a portion od life of the animal. The mass of the foot is invaded by a portion of viscera, at least the alimentary canal, liver superficially the gonads.

(iv) *Head.* A district head is absent in*Unio*. The eyes and tentacles are also absent. The large mouth opens beneath the anterior adductor muscle. It is boarded by a pair of broad, lamellar *labial palps* on each side.

Locomotion

The locomotion is performed by muscular foot. It is effected by a combination of blood pressure and muscle action. Engorgement of the foot with blood, coupled with the action of pair of *pedal protractor muscles,* produces extension.The water in mantle cavity and the blood act as a hydrostatic skeleton. The pressure of these two fluids is elevated by the adducted valves. Blood from the visceral mass is forced down into the pedal haemocoel, causing the foot to dilate and anchor into the substratum.

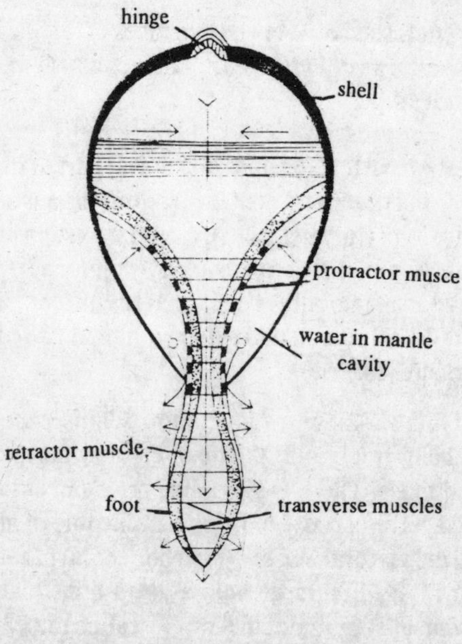

Fig. 5.4 *Unio.* Diagramatic, T.S.

Withdrawl of the foot is effected by the contraction of an anterior pair and a posterior pair pf retractors, which are also attached to the foot and shell. The scars of the pedal retractors are situated just above those of adductors.

Coleom.

Coelom is greatly reduced due to the development of connective tissue, unstriated muscle-fibres and blood-sinuses.Coelom is a haemocoel, filled with blood. It is represented by three small cavities: (a) a single ovoidal chamber, called pericardium, which is situated dorsally and in it, heart is located, (b) the cavities of the kidney or urocoel, (c) gonocoels or cavities of gonads. Originally, all the three were intercommunicated but due to gradual separation, the cavities of the gonads are completely separated, while the cavities of kidney and pericardium are inter-communicated.

Digestive System

The digestive system comprises of alimentary canal and a paired digestive gland.

I. *Alimentary canal.* The alimentary canal is complete. It is a coiled tube of varying diameter and starts from mouth and terminates into an anus. Buccal cavity, jaws radula and salivary glands are absent.

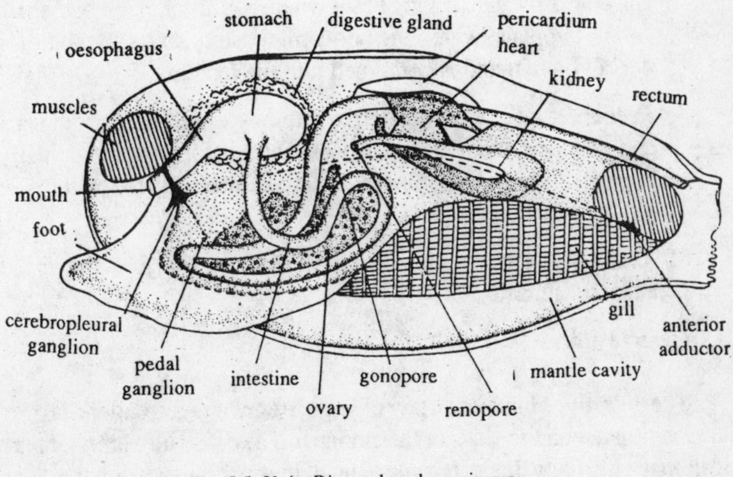

Fig. 5.5. *Unio* Disected to show viscera.

1. *Mouth.* The mouth lies at the anterior end of the visceral mass just below the posterior border of the anterior adductor muscle. It is bounded on either side by two soft, flat, triangular flaps called the *outer* and *inner labial palps.* The outer palps of the two sides unite above the mouth to from the *upper lip.* The inner palps of the two sides likewise fuse below the mouth to form *lower lip.* The outer surface of the palps is heavily ciliated. The mouth leads into the oesophagus.

2. *Oesophagus.* The oesophagus is a short, wide ciliated tube, which leads into stomach. The radula is absent as it is useless for the animal since it feeds on minute organisms.

3. *Stomach.* The stomach is a wide thick-walled rounded sac, situated dorsally in the visceral mass and is surrounded by paired digestive glands. The stomach is produced into a pouchlike outgrwoth known as *pyloric caecum.* It contains a trasparent gelatinous flexible rod called the crystalline style. It has proteinous matrix deposited with amlyse which is a carbohydrate splitting enzyme. The style rorates due to cilia of the stomach and brings about the mixing of enzymes with food. The crystalline style is secreted by the cells of the stomach and is contained in a style sac.

4. *Intestine.* From the posterior end of stomach arises the intestine and enters the substance of foot where it coils on itself and again ascent the place where the pericardial cacity starts. It then runs dorsally again and is continued into the rectum.

5. *Rectum.* It is the terminal part of intestine which starts from pericardium and runs for a short distance before opening finally through the anus in the cloacal chamber. Internally the rectum has a ridge, the *typhlosole.*

6. *Anus.* The anus is situated on an anal papilla just above the adductor muscle.

II.Digestive glands

The digestive glands are a pair of large irregular green or dark-brown masses situated on the sides of the stomach. They lead into anterior part of stomach by many ducts. It is made up of highly branched tubules, line by columnar epithelium of unequal height and containing basal nuclei.

The cells of these glands also swallow the food particles and break the solid particles in addition to the secretion of ensymes. These cells intracellularly digest the protein and fats, and absorb the carbohydrates,

Food and feeding. Freshwater mussel is a cliary feeder, feeding upon micro-organisms and organic debris which is filtered from the respiratory nutritive current of water drawn into mantle cavity. A continuous current of water is maintained mainly by the action of lateral cilia of gill filaments. As the water current enters the mantle cacity through the inhalent siphon it slows down and the solid food particles settle down over the gill surface. These are removed to the edge of mantle where they are pushed back and are expelled out. The lighter food particles get entangled into the mucus secreted by the gills. By the beating of the cilia of labial palps, the mucus is drwan towards mouth. The particles of food which cannot be digested are dropped down into the mantle cavity, whereas the digestible particles are pushed into the mouth along the deep ciliated grooves between the labial palps.

Digestion.

The digestion is *intracellula.r* Food from the stomach enters the digestive gland, the cells of which readily ingested and break down solid particles. Protein and fat digestion is exclusively intracellular, and the cells of gland also absorb carbohydrates. The only extracellular enzyme is the carbohydrate digesting enzyme (amylase) released in the stomach by the dissolution of the crystelline style. The style pouch is lined with cilia, the beating of which rorates the style and moves it forward with the result that its free end constantly rubs against a particular portion of the stomach wall and its head is worn away getting mixed with stomach content. Digested food diffuses from the gland into the blood for distribution.

Undigested food particles are sent back into the stomach, whence they pass into the intestine and them into the rectum by ciliary action. Function of intestine or rectum is not propely understood. They may absorb water from faecal matters. From the rectum the faeces is passed through the anus into the cloaca, which it leaves with the outgoing current through exhalant siphon.

Respiratory System

The respiratory organ in *Unio* are a pair of *gills* or *ctenida and mantle.*

1. Ctenidia or gills.

There are two gills that hang in the mantle cavity, one on either sid between the mantle and visceral mass. The gills are plate-like hence they are called *lemellibranchs.* Each gill has a central axis and plate like structure on either side. Hence the gills are called *bipectinate gills.*

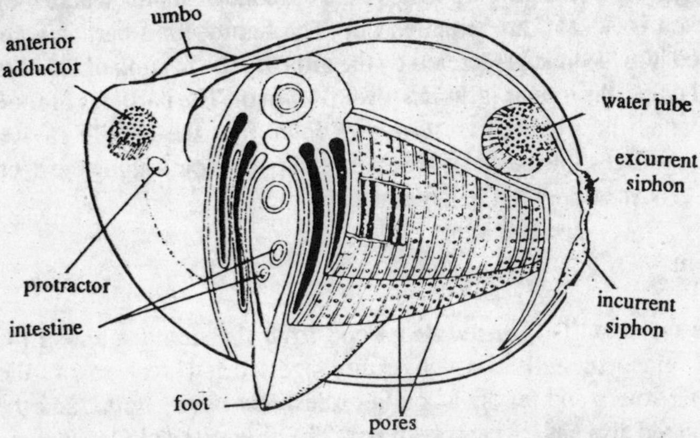

Fig. 5.6 *Unio* A schematic section of the mussel showing the respiratory organs. Water tubes have been exposed, whereas, the pores are highly magnified.

Each gill is a W- shaped structure. It consists of two palte-like structures called *laminae.* One lamina is situated on the inner side called the *inner lamina* and the other is situated on outer side and is called *outer lamina.* Each lamina is V-shaped structure and formed of two vertical plates called *lamellae.* The outer one is called *outer lamella* and the inner one is called *inner lamella.* The outer and inner lamellae are united together at their anterior, ventral and posterior edges, but are free dorsally. Each lamella is formed of a series of vertical filaments called *gill filaments.* The gill filaments of the two lamellae of a lamina are continuous at the free ventral side. The gill filaments are joined to one another by horizontal bars called *inter- filamentar junctions.* Two lamellae of a lamina are joined together by *inter-lamellar junctions.* The inter-lamellar junctions divide the cavity lying between the lamellae into vertical chambers called *water tubes.* Dorsally all the water tubes open into a large cavity called *supra-branchial chamber*. This chamber communicates to the exterior through the *exhalant siphon*. The portion of the mantle cavity lying below the gill is called *infrabranchial chamber* which communicates to the exterior through the inhalant siphon. The lamellae are perforated by minute openings, the *ostia*. They are located between the gills filaments and inter filamentar junctions. Through ostia, water tubes open into the mantle cavity.

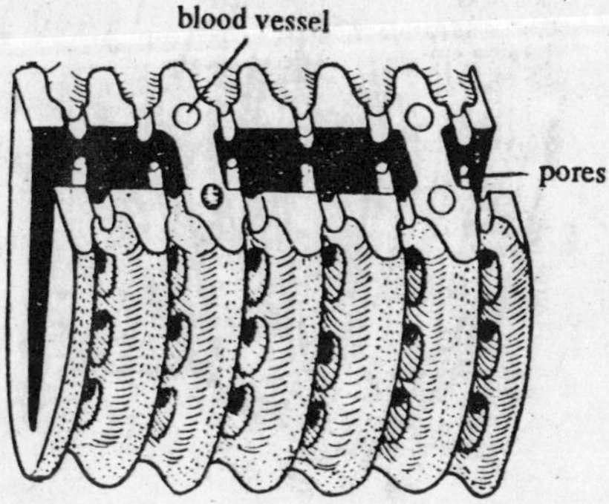

Fig. 5.7 *Unio*. Highly magnified gill

Each gill filament is covered by a layer of ciliated epithelial cells. Internally the gill filament is supported by chitinous rods.

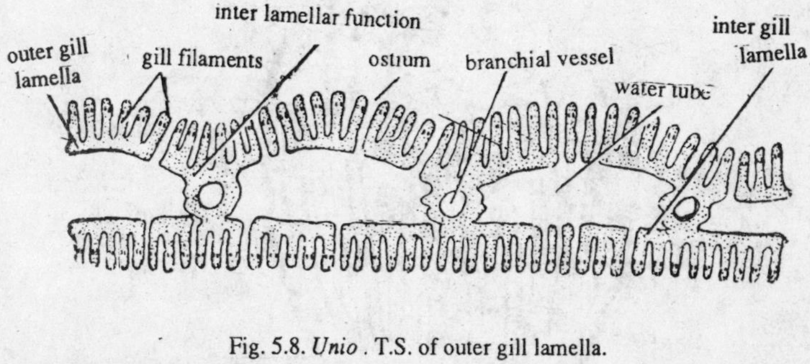

Fig. 5.8. *Unio* . T.S. of outer gill lamella.

Attachment of gills.

The mode of the attachment of the gills is very peculiar, as it determines the course of the water-current in the body. The outer lamella of the outer lamina is attached to the mantle along the whole length by

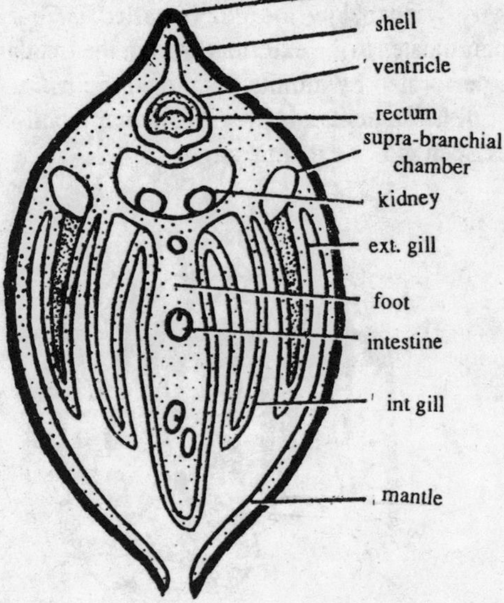

Fig. 5.9. T.S. of *Unio* through anterior region

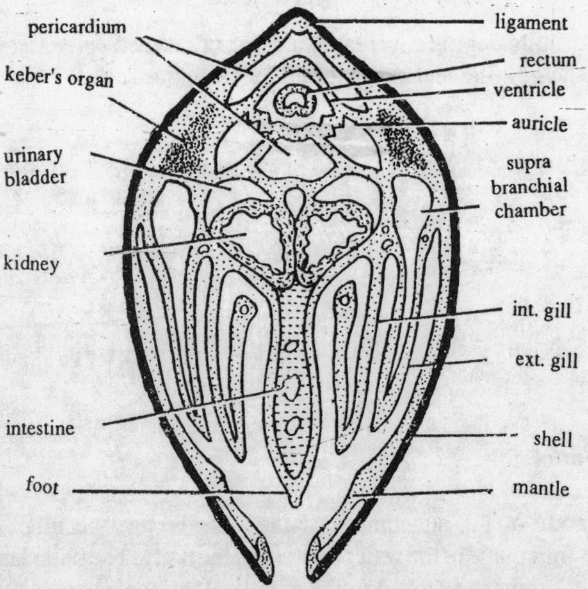

Fig. 5.10. T.S. of *Unio* through middle region.

its upper edge from anterior to the posterior end. Inner lamella of the outer lamina and the outer lamella of the inner lamina are attached to the sides of the visceral mass a little below the origin of the mantle by dorsal edges. The inner lamella of the inner lamina is attched to the visceral mass in the anterior region, but is free in posterior region. Behind the posterior margin of the visceral mass inner lamella of inner lmina is also attached to its fellow of the opposite side. The inner lamina are attached together by their dorsal edge to the *posterior adductor muscles* on the sides of the visceral mass a little below the origin of the mantle. The outer lamella of the outer lamina is attached to the whole length of the mantle. The cavity between the upper edge of gill-lamella and the upper part of the mantle is called supra-branchial chamber, which continues posteriorly with the exhalent siphon.

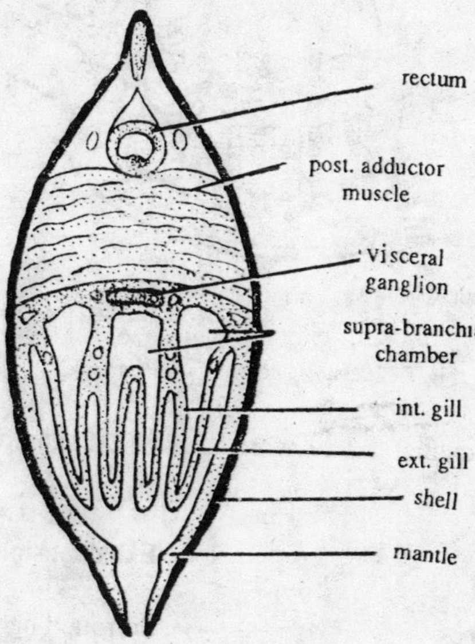

Fig. 5.11. T.S. of *Unio* through posterior region.

Blood supply

The gill receives venous blood from the kidney by the *afferent branchial vessels* and the oxygenated blood is conveyed to the heart by *efferent branchial vessels.*

Course of Water current

Beating of cilia of the gill filaments draws a continuous current of water into the infra-branchial chamber through the inhalent siphon. From here it enters the gills through ostia. Indise the gill, the water fills in the watertubes and flows out into the *supra-branchial chamber* are dorsal to each gill lamina and open posteriorly into the cloacal chamber. The water finally leaves the cloacal chamber through exhalent siphon which discharges water to the exterior.

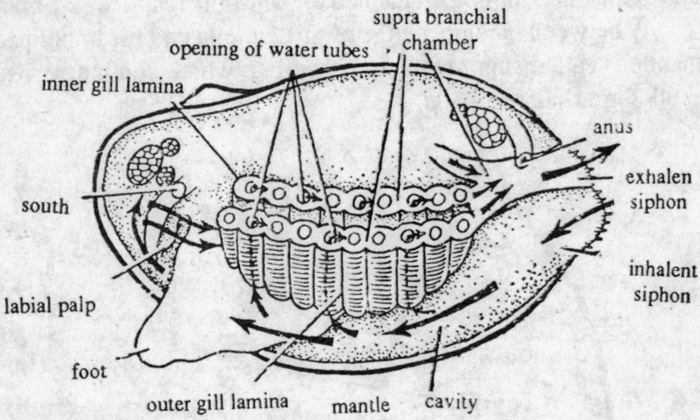

Fig. 5.12. *Unio*. Course of water current

Supra-branchial chamber → clacal chamber

↓

Exhalent siphon

↓

Surrounding water

↓

Inhalent siphon → Sub-branchial or infra-branchial chamber → Ostia → Water tubes (inside Gills) → Supra-branchial chamber

Circulatory system

The circulatroy system of *Unio* is said to be *open type* because the capillaries are absent. The circulatory system comprises three parts.

1. Heart
2. Blood Vessles
3. Blood

1. *Heart.* The heart is situated dorsally near the posterior adductor muscle. It is enclosed in *pericardium.* The pericardium is a thin-walled triangular chamber. It is lined by epithelium and is filled with a fluid and represents a part of coelom. It is communicated with the supra-bracnchial chamber through the kidneys.

The heart is three chambered, having two *auricles or atria* and a *ventricle.* The auricles are thin-walled, highly distensible, triangular chambers one on either side of ventricle. They are attached by their broad bases to the pericardium and open dorsally into the ventricle. Each auriculo-ventricular *aperture* in provided with a valve that opens towards the ventricle and prevents the backward flow of blood. The ventricle has the form of a horizontal tube and surrounds a part of rectum.

The heart consists of smooth muscle-fibres covered externally by an epithelium called the *epicardium.* Endocardium is absent. The heart beats 20-100 times per minutes.

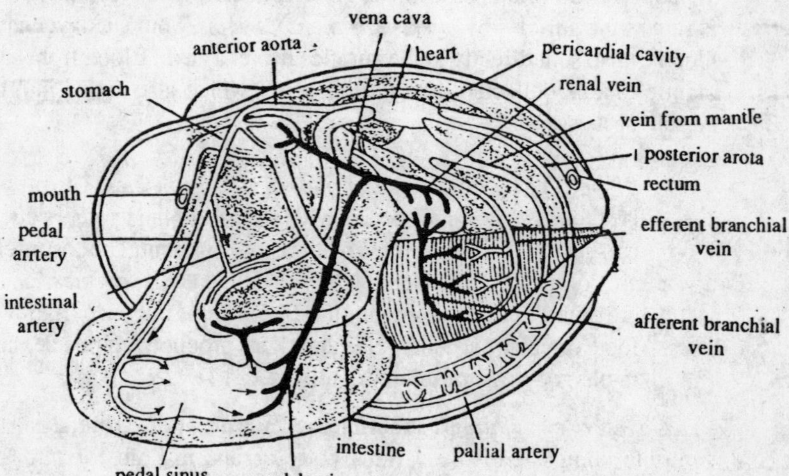

Fig. 5.13 *Unio* circulatory System

2. *Blood vessels*. The blood vessels comprises the (i) *arteries (ii)* sinuses (iii) veins,

(i) *Arteries*. From each end of ventricle arises, the *anterior aorta* passing above and *posterior aorta* below the retum. The anterior aorta supplies to the anterior parts of the body i.e. mantle, foot and viceral mass by *anterior pallial, pedal and visceral* arteries. The posterior aorta supplies to the posterior parts of the body i.e. rectum, mantle, nephidia, pericardium etc. by *rectal, posterior pallial, nephridia, pericardial* arteries.

(ii) *Sinuses*. The arteries ramify all the organs and finally empty into sinuses. From small sinuses blood is returned by larger sinues. The sinuses lack the epithetical lining of true blood vessels and connect directly with veins.

(iii) *Veins*. Blood from different parts of body is collected by many small veins in the end it reaches the large view called *vena cava* which is placed longitudinally in the body. The vena cava is situated below the pericardium and between the kidneys. The blood from vena cava is send to kidneys by the *renal vein*, here nitrogenous wastes are removed. Blood from kidneys is sent to the gills for aeration by *afferent branchial vein*. Which gives off branches to all the gill-filaments. The oxygenated blood is sent ot the auricles by *efferent branchaial veins*. Some deoxygenated blood is also sent directly to the auricles by vena cava. Blood from each mantle lobe is collected by a *pallial vein*, which also opens into the auricle of its side.

3. *Blood*. The blood consists of *plasma* and *corpuscles*. The plasma is colourless in some species where as faint blue in others due to presence of *halmocyanin*, a respirastory pigment. The cospuscles are colourless and amoeboid hence called the *leucocytes*. Some are granular other non-granular. Blood helps in transportation of food. oxygen, carbondioxide and waste products, corpuscles help in phagocytosis of microorganisms.

Course of circulation The auricles receive the purified blood and the ventricle pumps it to the various body organs in both forward and backward directions. The various veins and sinuses collect the blood from body organs and carry it to the kidney and gills for purification. The course of blood is again carried to the auricles. The course of blood circulation can be shown as under:

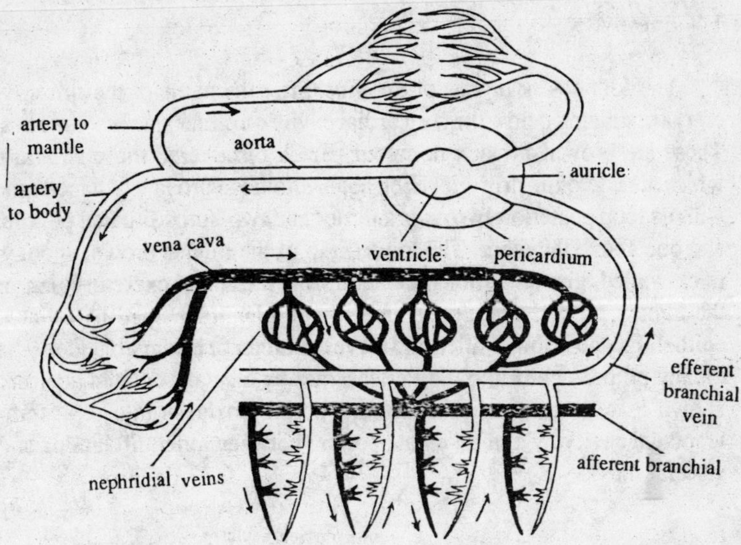

Fig. 5.14 *Unio* Circulatory System diagrametically.

Pallial Arterty → Aorta → Blood → capillaries → Smaller veins and sinuses → Vena cava → Renal Veins → Kidneys → Afferent branchial vessels → Gills → Efferemt branchial vessels → Auricles speaee

Pallial → Mantle → Pallial veins → Auricles speaee

Arterty → Ventricle → Auricles speaee

Excretory system

In *Unio* followings are the excretrory organs.

1. A pair of kidneys or organs of Bojanus
2. Keber's organ

1.Kidneys

The kidneys or organs of *Bojanus* (after the name of their discoverer) are situated below the pericardial cavity on either side of vena cava. These are brownish-black in colour. Black discovered these structure which are coelomic in origin. Each nephridium is a broad, U-shaped tube with its loop posterior, two ends anterior and two limbs situated purallel and one above the other. The lower limb of the tube is brown, spongy, thick walled, glandular in nature and forms the actual excretory part or the kidney proper. The upper limb is non-glandular lined by ciliated epithelium and is thin-walled and serves as ureter or urinary bladder. The kidney proper opens into the Pericardium by a small, ciliated aperture known as reno-pericardial aperture. The ureter opens into the supra-branchial cavity by a small renal aperture between inner gill lamina and visceral mass.

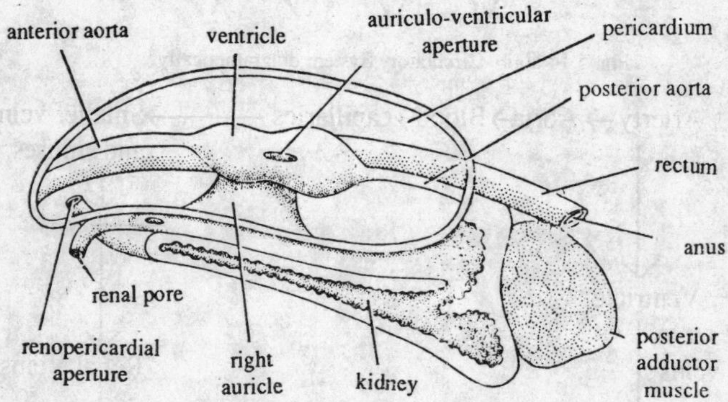

Fig. 5.15. *Unio*. Excretory System.

The ventral glandular part of the kidney extracts nitrogenous products and guanin from the pericardial fluid and the blood. The ciliated epithelial lining of the bladder creates an outgoing current, which carries excretory fluid to the outside.

2. Keber's Organ

Keber's organ is a large reddish-brown gland situated in front of pericardium. It discharges waste products in the pericardium, whence it is eleminated via kidneys.

Nervous system

Due to sluggish, sedentary life, the nervous system of *Unio* is greatly reduced. It consists of three pairs of gangtia, connnectives and commissures. These are:

1. Cerebro-pleural ganglia

At the base of the labial palps, just outside the corners of the mouth on either side, is placed a small yellowish, somewhat triangular *cerebro-pleural ganglion*. Each is formed by the fusion of two ganglia a cerebral & a pleural. The ganglia of both the sides are equivalent of the brain. They are connected with each other by a thin transverse cerebral commissure, which passes over the oesophagus. Each cerefro-pleural ganglion gives out - (i) an *anterior adductor nerve* to the corresponding muscle, (ii) a *labial nerve* to the labial palp, and (iii) an *an anterior pallial nerve* to the anterior part of the mantle. Besides, each ganglion gives off two conspicuous connectives, a *cerebro visceral connective* which runs posteriorly to unite with the viseral ganglion of that side, and a *cerebro-pedal connective* which passes ventrally into the foot to unite with the pedal ganglion of the same side.

2. Pedal ganglia

The pedal ganglia are situated in the foot about 1/3 from its anterior end. They are present above muscular part and below the visceral mass.

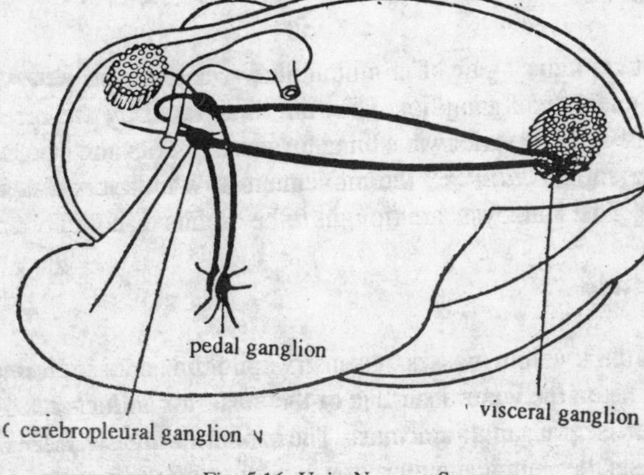

Fig. 5.16. *Unio*. Nervous System.

Both the pedal ganglia are closely associated so as to form a bilobed mass which supplies the nerves to the foot, its muscles and the statocysts. Each pedal ganglion is connected to the cerebro pleural ganglion of its side by the cerebro-pedal connective.

3. Visceral ganglia

A pair of visceral ganglia are situated mid--ventrally on the posteriar adductor muscles. The visceral ganglia are fused together to form a flattened, somewhat rectangular mass. On each side, the visceral ganglionic mass gives off (i) a dorsal pallial nerve, (ii) a posterior pallial nerve to the posterior part of the mantle, (iii) a posterior renal nerve to the kidney, (iv) branchial nerve to the gill. It is connected with the cerebro-pleural ganglion of its side by a thin cerebro-visceral connective which gives off several small nerves to the visceral mass.

Autonomous nervous system. Recently, *E.I.Thomas* (1974) has worked out anatomy of *Lamellidens* and has reported presence of minor ganglia in the chief muscles and has suggested that they are autonomous.

Sense Organs

The sense organs are pooly developed due to sedentary mode of life. The main sensory structures are (i) Statocysts (ii) Ospharidium and (iii) Sensory cells.

1. Statocysts

The foot contains a pair of minute hollow vesicles, the *statocysts*, one close to each pedal ganglion. They are innervated by the cerebro-pedal connectives. The statocyst is lined by sensory cells and contains a mass of lime, called *statolith*, the movement of which stimulates the sensory cells. The statocysts are thought to be organs of equilibrium.

(ii) Ospharidium

A pair of dark yellow masses of sensory epithelial cells are found at the base of gills on the ventral surface of the posterior adductor muscle and above the visceral ganglionic mass. The ospheridium is a *chaemoreceptor* as it tests the chemical nature of water entering the mantle cavity by way of the inhalant siphon.

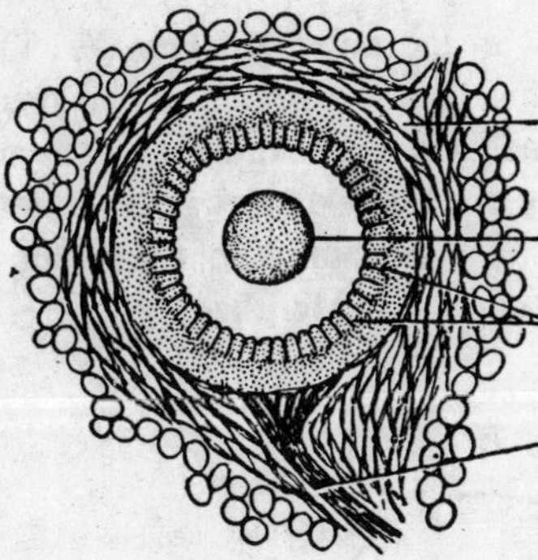

Fig. 5.17 *Unio*. Statocyst.

(iii) Sensory cells.

There are also other light responding devices found in the edges of mantle lobe and abundant on the inhalant siphon. They are regarded to be tactile organs and are also sensitive to light.

Reproductive System

The *Unio* is *dioecious* i.e. sexes are separate. The reproductive organs consist of the gonads and their ducts.

Male reproductive organs. These consist of a pair of testis which are white-coloured, highly branched, glandular masses situated in the visceral mass between the loops of intestine. From each testis a *vas deferens* arises which opens into the supra-branchial chambers by male genital pore situated in front of renel aperture.

Female reproductive organs. These are a pair of pink coloured glandular mass of branched tubules occupying the same position as testes in male. These become greatly enlarged during breeding season. Each ovary gives out an oviduct which opens by female genital aperture situated in front of renal opening. The accessory reproductive organs are absent.

Fertilization

The eggs are fertilized in the outer gill laminae. The eggs are collected in the supra-branchial chamber from here they reach the water tubes of the outer gill-laminae and are held there by the mucous. The sperms come out of the body of male body along with outgoing current of water. Some of these reach the gill laminae of the female through the incurrent siphon with water current and fertilize the eggs.

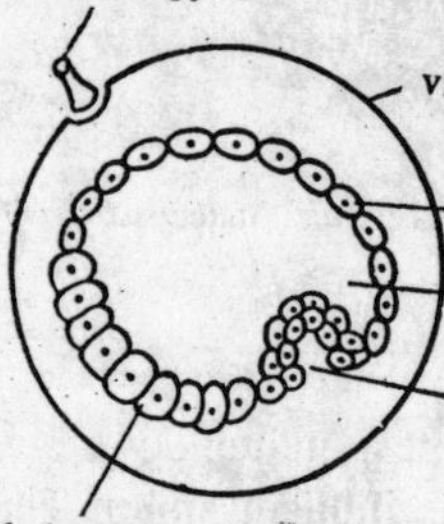

Fig. 5.18 *Unio*. A gastrula

Development

The fertilized ovum undergoes segmentation. The cleavage is complete (holoblastic) but unequal. A hollow, spherical embryo, the *blastula* with a fluid filled cavity the blastocoel is formed. The bigger blastomeres are called *macromeres* and the smaller ones, *micromeres*. In gastrulation some of the macromeres are invaginated, resulting in the formation of an exceedingly small cavity, the *archenteron*. Later on there is an extensive invagination of macromeres which very greatly reduces the blastocoelic cavity. This is the rudiment of *shell gland*. Some large cells are budded off into the blastocoel, where they aggregate to form *mesoderm*. The invagination of the macromeres which constitute the rudiment of the shell gland, marks the dorsal surface of embryo, while the posterior end is marked by a tuft of cilia. The cells of the shell gland, which are tall, columnar, secrete an unpaired shell. Later arises a shell consisting of two valves, which are more or less triangular in appearence and have their ventral ends sharply incurved to form *hooks* that bear *spines*. The mass of cells consisting the embryo is divided into two on the ventral aspect, but remains undivided dorsally. The two ventral halves are the rudiments of mantle lobes. These bear peculiar brush-like sense

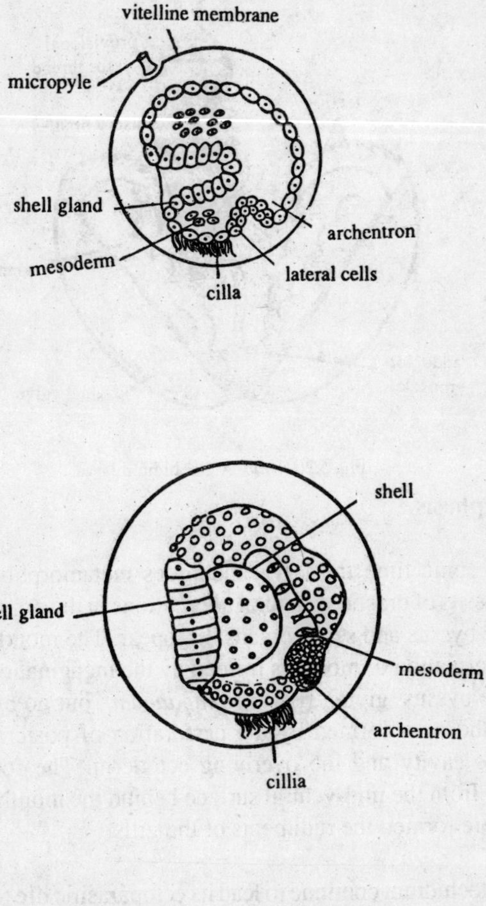

Fig. 5.19. *Unio* A early and late embryo.

organs. A long, thread-like adhesive organ, the *provisional byssus* appear on the ventral surface of the body. The mesoderm produces a large *adductor muscle* that stratches between the two valves of the shell. While in the lamina, which serves as a *brood pouch* the larva, called *glochidium,* is nourished by the secretions of laminar wall but on escaping from the body of mother through the exhalent siphon, it starts leading an ectoparasitic life on the body of some fish, which is persisted till its metamorphosis into adult. The glochidium is a minute larva 0.1 - 0.5 mm wide.

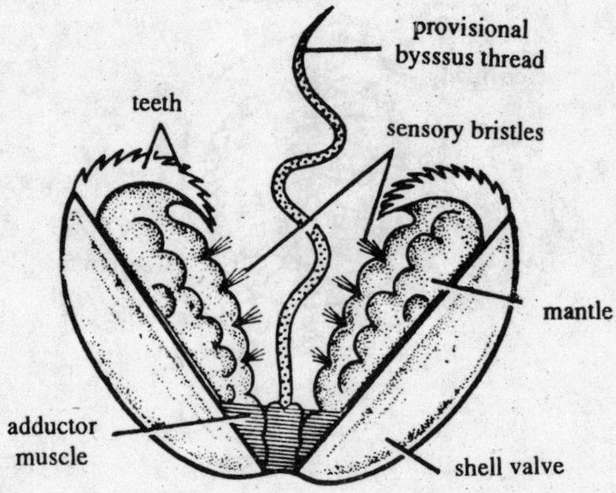

Fig. 5.20. *Unio.* A glochidium larva.

Metamorphosis.

After some time the larva undergoes metamorphosis, which, as usual, consists of drastic structural alternations in the organisation of the body. The byssus and sense organs disappear. The mouth and anus are formed. Formation of mouth is formed by the ingagination of ectoderm behind the byssus giving rise to *stomodaeum,* but no *proctodaeum* is formerd, the *anus* formed by the perforation of posterior end of the archenteric cavity and the overlying ectoderm. The foot arises as a projection from the mid-ventral surface behind the mouth and on either side of it are formed the rudiments of the gills,

The glochidium continue to lead its ectoparasitic life to the end of its larval period, after the completion of metamosphosis it leaves the host and begins its independent life, growing on the adult state. The time taken by larva to metamorphose into little clams is about 10 weeks.

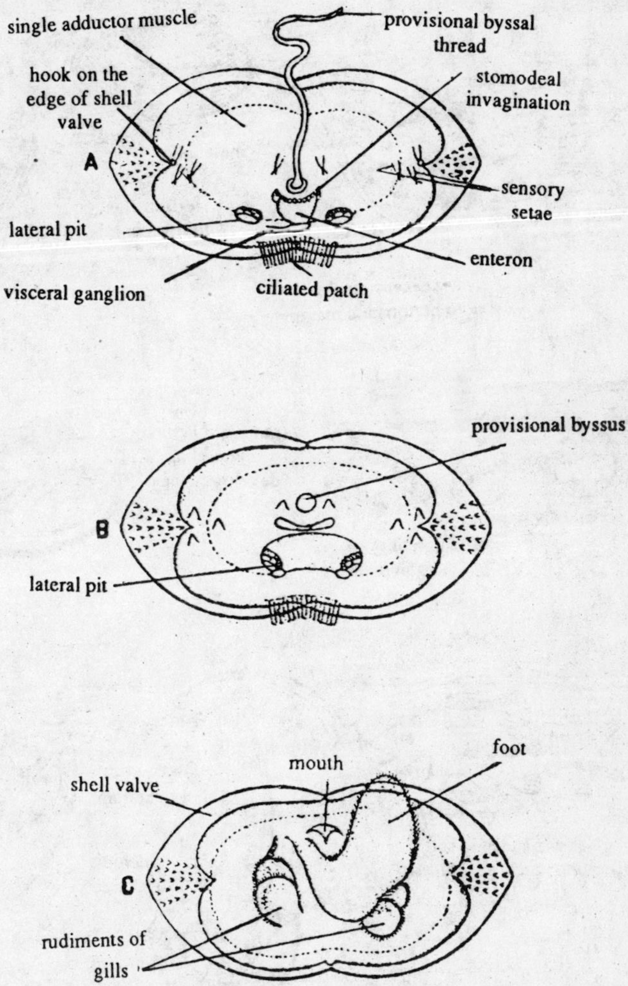

Fig. 5.21 *Unio*. There stages in metamorphosis

Significance of glochidium

Glochidium is an ectoparasitic larval stage on a fish host. It helps in dispersal, protection and means of nourishment.

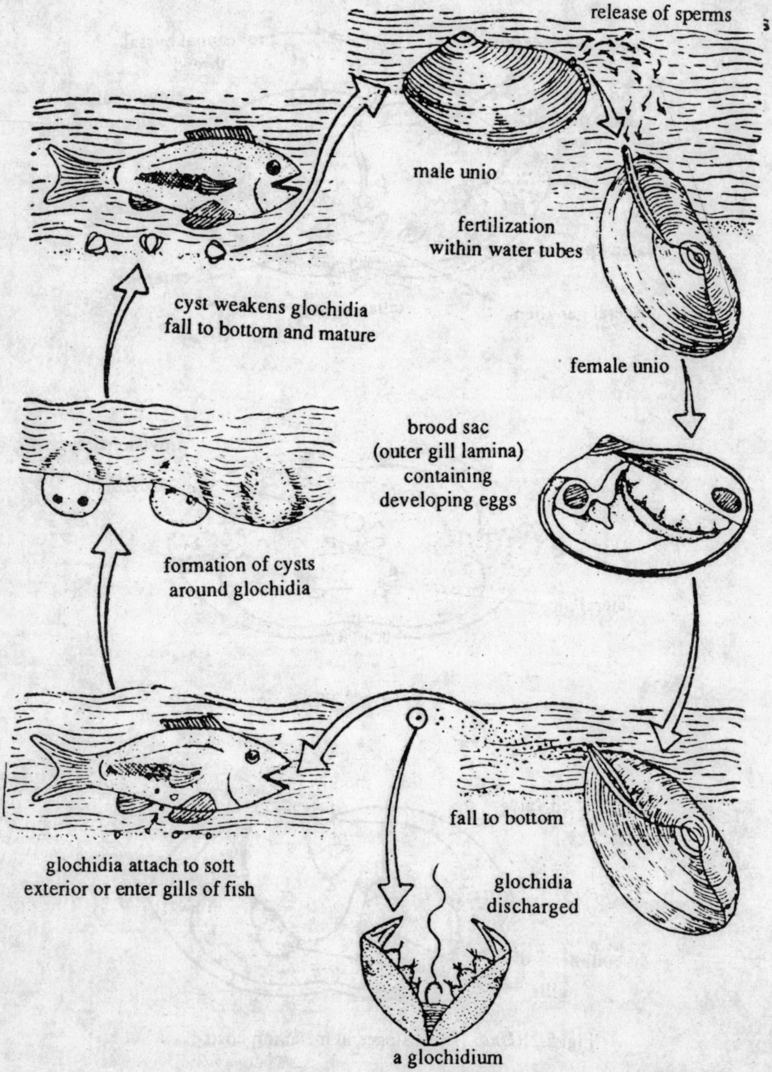

Fig. 5.22 *Unio*. Life cycle

6

SEPIA

Sepia comes under class Cephalopoda in which the foot is trasnformed into circumoral appendages completely surrounding the head & the epipodium is modified term as *funned or siphon or exhalent muscular tube*.

SYSTEMATIC POSITION

Phylum	-	Mollusca
Class	-	Cephalopoda
Sub class	-	Dibranchiata
Order	-	Decapoda
Genus	-	*Sepia*

Distribution

Sepia is commonly called cuttle fish. It is cosmopolitan in its distribution and is more common in warmer seas like Mediterranean, Indian ocean, Pacific etc.

Habits and habitats

Sepia is a marine mollusc. It inhibits shallow waters usually living between stones and in crevices of rocks but often burrows in sand,

It is a good swimmer. The swimming movements are forward and backwards by the fins and funnels. It feeds on crustaceans and fishes. It is capable of colour changing. When followed by an enemy, it discharges into water ink from the ink gland. Sexes are separate Breeding occurs in summer.

External features

Shape and size. Sepia has a fishy, bilaterally symmetrical and dorso-ventrally flatenned body. The anterior and posterior ends of the body, infact, represent the dorsal and ventral ends due to much elongation of dorso-ventral axis of the body. The body-colour is brownsih and it is about 20 cm. in length.

The body is divisible into three parts :

1. Head
2. Neck
3. Trunk

1. *Head* ‹The head is almost globular in form. It bears a pair of large, highly developed *eyes* at the sides and *mouth* at its free end. The mouth is surrounded by five pairs of muscular processes. These are differentiated into four pairs of short and stout *arms* and one pair of long *tentacles* retractile into large pits at their bases. When extended, they are about 2/3 of the length of the body. They have strongly convex outer surface and flat inner surface. The inner surface bears four longitudinal rows of *suckers*. Each sucker is a muscular, shallow *cup* with a narrow, horny rim and supported on a short, thick *stalk*. The tentacles bear sucker on their expanded club-like tips only. The tentacles are used in the capture of prey and in copulation. The left arm of the fifth (posterior) pair is slightly modified in the male. The modification simply pertains to the loss of some of the suckers. This arm is called as *hectocotylized* and is used for transferring spermatophores into the mantle cavity of female. The arms, tentacles and the head proper are collectively called cephalopodium Some workers regard arms as the main parts of the foot

However, it has been suggested that arms should not be described as modified parts of foot but as cephalic appendages. According to this view, the *funnel* or *siphon* should be regarded as the molluscan foot,

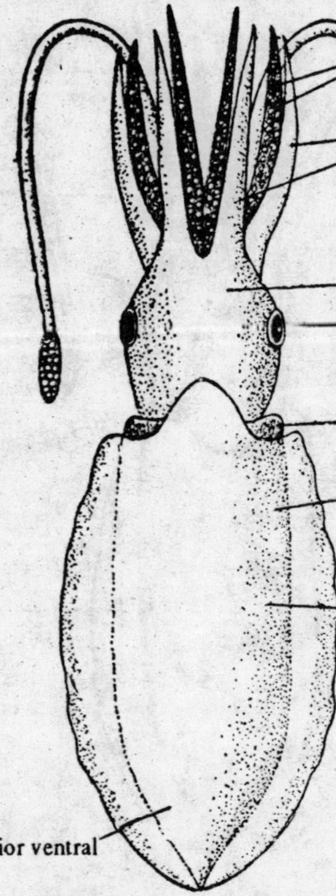

Fig. 6.1 *Sepia*. Dorsal view.

hence the name *Siphonopoda* has been suggested by some authors. The funnel is situated on the posterio - ventral side behind the head and is regarded a modified part of foot.

2. *Neck* It connects the head with trunk. It bears a funnel which opens to out side by a narrow apertuse and by a large apertuse into mantle cavity.

3. *Trunk* The trunk is elongated somewhat flatenned and shield-shaped. Its braod base is directed orally and the pointed apex aboally. It is bordered by a narrow *fin* on either side which helps in locomotion at leisure. One side of trunk is darker, slightly convex and hard due to the presence of shell inside.

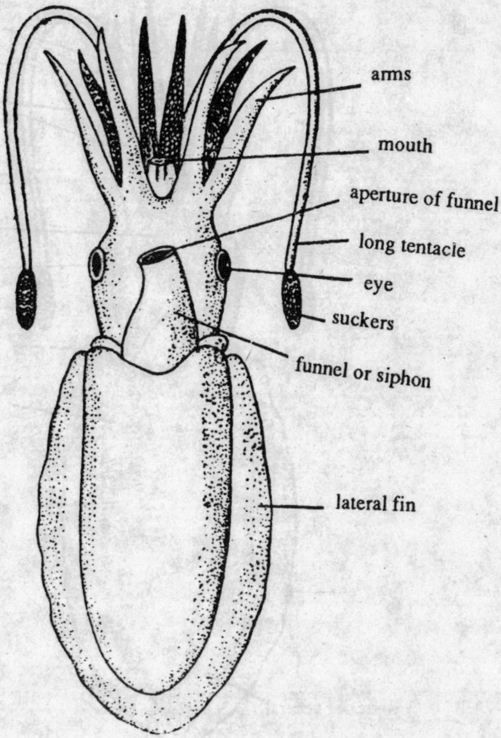

Fig. 6.2 *Sepia*. Ventral view.

Mantle

The trunk is covered by a thick fold of mantle, which terminates towards the oral surface in a ridge around the neck. The head can be partially retracted. Posteriorly the ridge forms a large cavity bounded by the mantle called the *mantle* cacity, later contains the entire viscera.

Funnel

Below the head lies a large conical muscular tube, the *funnel or siphon* projecting beyond the neck. It internally communicates with the mantle cavity by a wide aperture, thus the funnel is the main out-let of the mantle cavity. A pair of cartilaginous knobs on the mantle fits into corresponding sockets on the posterior ventral surface of the funnel. The funnel does not only carry to the exterior the foul water of respiration, the faccal matters and the products of excretory and reproductive organs but also takes an important part in locomotion.

Skin

Flat contractile, pigment-containing cells, the *chromatophores* occur in the deeper layers of the integument over the entire surface. They have elastic membrane from which bundlcs of muscle-fibres radiate to the surrounding tissue. The chromatophores are of three kinds reddish, yellowish-brown and orange. They can be detailed by the contraction of rediating muscles. The living *Sepia* can change its colour Besides the chromatophpores, a number of cells, the *iridocytes* are present in the integument which cause the peculiar iridescence in addition to the vivid colouration caused by the chromatophors.

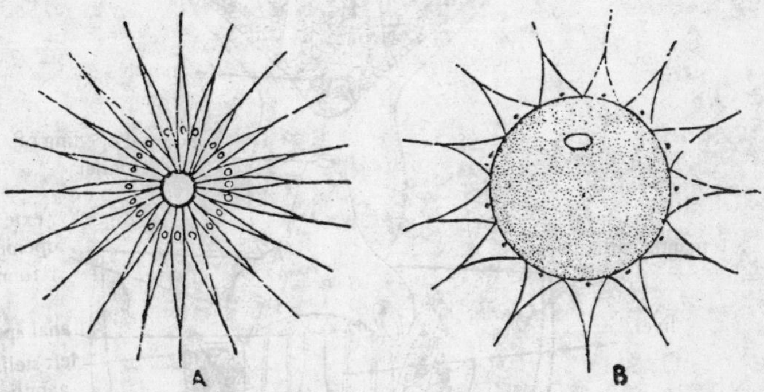

Fig. 6.3 . *Sepia*. Chromataphore A = contracted B = expanded

Shell

The shell is internal and is called the *cuttlebone* or *sepiostaire*. It is enclosed in a sac of mantle on the anterior surface of the trunk. It is secreted by the sac itself. It is flat and roughly oval in form with rounded oral end the postracum and a pointed aboral end, which has a sharp *spine* or *rostrum*.

The shell is non-living and consists of a large number of closely arranged calcareous laminae. The laminae enclose air-spaces between them to make the shell-light. Thus the shell besides acting as endoskeleton also serves as a *float* and help to maintain equilibrium of the body being dorsal in position. The shells of dead *Sepia* are often found on the sea-coast thrown by the tides. These are often called the *sea-foam.*

Mantle cavity

It is a large space enclosed between the trunk and the mantle on the ventral side. It is closed on all the sides except the oral side, where it communicates with the exterior by two passages arround the neck, the *pallial aperture* and a large tube-like *funnel* or *siphon*. The funnel has a wide *internal opening* and a narrow *external opening*. Basally, on its ventral

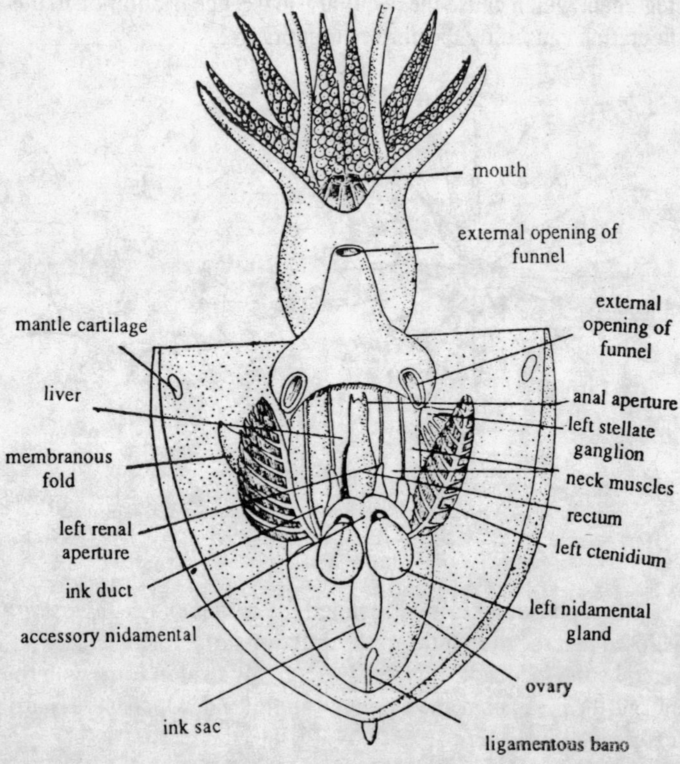

Fig. 6.4. *Sepia*. Showing Ciscera

surface, the funnel bears a pair of cartilaginous sockets, called the *funnel* or *infundibular cartilages.* Opposite them, on the internal surface of the reflected mantle folds, are seen a pair of oval cartilaginous knobs, the *mantle cartilages* . Similarly, the dorsal surface of neck carries the *nuchal cartilage,* which fits aganist the *dorsal cartilage* on the mantle. By enlarging the mantle cavity, the *Sepia* draws waterinto it through the pallial aperture. When the mantle cavity is reduced, the mantle and infunfibular cartilages get interlocked to close the pallial aperture. There fore the water leaves the mantle cavity through the funnel. The funnel is guarded by a flap-like *valve* that opens out wards.

The whole of the mantle cavity is occupied by visceral mass having various organs. The digestive gland is situated anteriorly on either side of the retractor muscles of head and funnel. There is a pair of plume-shaped ctenidia situated posteriorly in the mantle cavity. Between the ctenidia is found anus. On each side of the anus there are thin openings of renal sac called renal poles or apertures situated on renal papilla. On the left hand side is the opening of genital aperture also situated on a papilla, penis is in the male and oviduct is found in female. On the mantle wall where neck joins the trunk are situated two large stellate ganglia one on each side. Posteriorly behind the nidamental glands is found an ink-sac which contains ink made of melanin pigment. The duct of the ink-sac, runs forward ventral to the rectum and opens it dorsally close to the anus. In male *Sepia,* testis is situated covered by an ink-sac while in female renal sac are invisible as they are hidden by nidamental glands and a pair of accessory inidamental glands.

Locomotion

Sepia exhibits two types of locomotion. They are :

(i) By fins
(ii) By the funnel

(i) *By fins*. The fins which are fringed along the lateral margins of body. The fins bring about swimming by their undulations.

(ii) *By the funnel*. The funnel or siphon brings about backward darking movements. During this movement, the mantle cavity is filled with water. When the mantle contracts, the pallial apertuse closes due to interlocking of the mantle and infundibular cartilages and this expels the water from the mantle cavity through the funnel as a jet. As the jet of water is forced forward, the animal is pushed backward. The animal can dart forward also by bending the siphon backward

Coelom

It is a pouch being divided into *oral* and *aboral* parts by a constriction. The oral is the anterior part or *pericardium* in which heart is enclosed. It communicates with the cavities of the kidneys or renal sac by a pair of *renopericardial* apertures. The aboral part forms the gonocoel which encloses the gonads.

Digestive System

The digestive system consists of
1. Alimentary canal
2. Digestive glands.

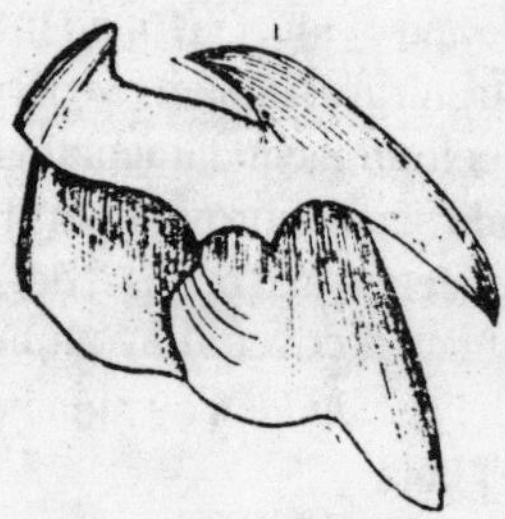

Fig. 6.5. *Sepia*. A Jaw.

1. *Alimentary Canal* : The alimentary canal starts from mouth and terminates into anus. *Mouth* is present among the oral arms and is surrounded by a thin, fleshy circular *lips* beset with numerous papillae. Two *jaws* are located in the lips. The jaws have somewhat the appearence of the beak of a parrot. The month leads into a large thick-walled, muscular *buccal cavity* or *pharynx*. Buccal cavity contains on its floor the *odontophore* and the *radula* bearing numerous minute, horny *teeth*. The buccal cavity leads into the *oesophagus*, which is a long narrow tube. It runs straight towards the aboral end, passing between two lobes of liver. The oesophegus opens into a rounded thick walled muscular bag, the

stomach. The stomach is followed by the tubular *intestine*. At the junction of stomach and intestine, a wide, curved sac opens into the alimentary canal. It is known as *caecum*. The intestine runs forward into a *rectum*. The rectum opens into the mantle cavity by *anus*. A pair of leaf-like *anal-valves* project at the sides of anus. Their function is uncertain.

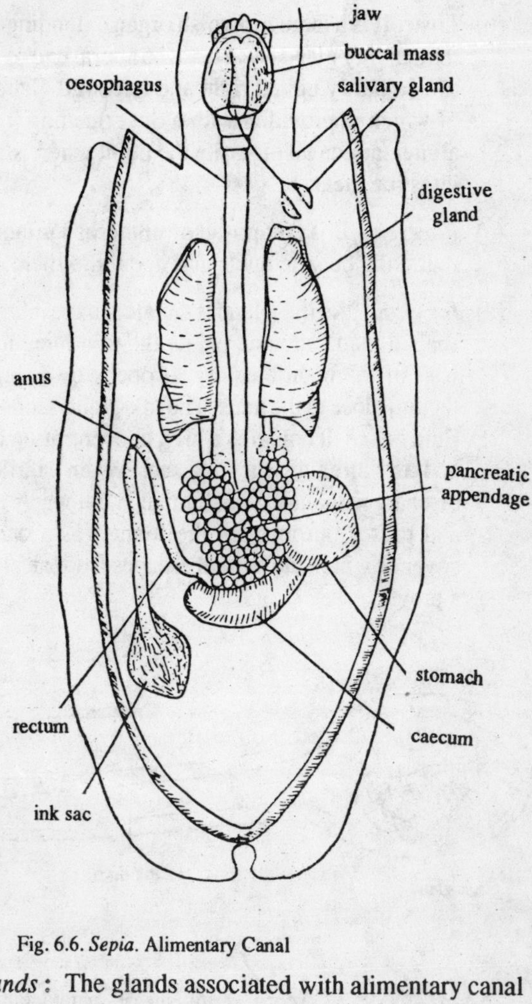

Fig. 6.6. *Sepia*. Alimentary Canal

2. *Digestive glands* : The glands associated with alimentary canal are:

(a) *Salivary glands.* A paired salivary glands are present in the head behind the cranial cartilage. A *salivary duct* leds inwards from each gland and joins its fellow from the opposite side to form *medium salivary duct* which opens into buccal cavity. The function of salivary glands related to saliva has not been misnamed because their secretion is used to paralyse the prey hence these should be called *poison glands.*

(b) *Liver.* It is a large brownish organ extending from the vicinity of the salivary glands nearly to the aboral end of the body. It consists of two partly united right and left solid glandular portions, each of which is provided with a duct opening into the cavity of the alimentary canal opposite the point where stomach, caecum and intestine meet.

(c) *Pancreas.* It is composed of numerous minute vesicles that surround the hepatic ducts and open into them.

(d) *Ink gland.* An ink-gland is situated in the ink-sac. Ink-sac is a pear shaped gland situated below the thin integument of the anterior wall of the mantle cavity. It opens by a duct dorsally into the rectum close to the anus. The ink-gland secretes a brown or black fluid or *ink.* It contains a high concentration of melanin pigment and is stored in the ink sac. When startled, the cuttle fish discharges the ink, which mixes with water in the mantle cavity and passes out through the funnerl as a dark-cloud, under the cover of which the animal escapes from an enemy or approaches a prey.

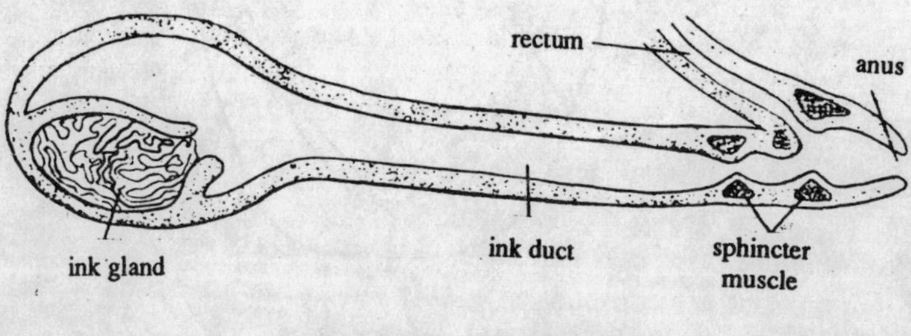

Fig. 6.7 *Sepia.* L.S. of ink gland

Food. Food of *Sepia* consists of crustaceans e.g., Prawn and Shrimps; molluscs and small fishes. Prey is caught by the protrusion of the tentacles. The prey is atached to the tentacles by the suckers and drawn in within the reach of the arm. Jaws which are situated in the mouth help in crushing the food, radula in grasping the flesh and in swallowing. A pair of salivary glands is also situated in *Sepia.* These glands produce a posion which kills the prey.

Digestion. The food caught by tentacles is paralysed by salivary glands crushed into pieces by jaws swallowed with the help of radula, reaches the stomach where it is mixed with the digestive fluids from the liver and pancreas. The digestion is complcted in the caecum where liquid products are absorbed while undigested food is passed into the intestine where some absorption may take place. The undigested food material thrown out of the anus. The hepato -pancreas secretes digestive enzymes and serves for the absorption and storage of the food and removal of waste products. The food stored in the liver is not directly absorbed but received through the blood.

Respiratory System

The respiratory organs are paired, large, plume-shaped *ctenidia or gills,* lying in the mantle cavity, one on either lateral sides. Each ctenidium is bipinnste, with numerous delicate *lamella*e on either side of a central axis. The respiratory surface of the lamellae is increased due to the presence of various folds on the lamellae. Internally the lamellae are not competely in contact with each other. Each gill receives venous blood through an *afferent branchial vessel* from the branchial heart of its side. The blood passes in a system of minute branches through the lamellae and is finally collected into an *efferent branchial vessel* leading to the auricle.

Respiratory mechanism

The muscular mantle rhythmically contracts and expands, thus the mantle cavity alternately increases and decreases. When the mantle cavity enlarges, fresh water flows into it through the pallial apertuse. This water passes over the ctenidia. Exchange of gases occur in the vascular lamelle. Carbondioxide is diffused out in the water and when the mantle contracts, the water is expelled out through the funnel.

Circulatory system.

The circulatory system is well developed with a complete separation of venous and arterial blood. It consists of heart, blood vessels and a system of capillaries and blood.

Heart. There are three hearts; a large systemic or arterial heart and a pair of small branchial or venous hearts.

The systemic or arterial heart lies in the middle of the visceral mass enclosed in the *pericardium.* It is three chambered, having a large, median, thick-walled *ventricle* and two small lateral, thin-walled *auricles.* The ventricle is divided into two lobes by a constriction. It supplies arterial blood through a large *oral* or *cephalic aorta* and a small *aboral* or *posterior aorta* to the anterior and posterior regions of body. The aurcles receives oxygenated blood by the *efferent branchial veins.* This blood passes into the ventricle and is send out via aortae.

The branchial hearts lie one at the base od each ctenidum. They are simple contractile sacs that receive and supply deoxygenated blood by *afferent branchial veins.* A rounded glandular body called *appendage* is attached to each branchial heart.

Blood vessels. The oral and aborae aortal divides into small branches, the *arteries,* which divide and redivide to form very fine, microscopic vessels, the *cepillaries.* The cepillaries reunite to form large vessels, the veins. A large median *vena cava* returns blood from the head region. Infront of the rectum, it bifurcates into *right* and *left afferent* branchial veins, which run through the corresponding renal organs to the base of ctenidia. Here they dilate to form branchial hearts.

A pallial vein from the mantle and an abdominal vein from the posterior region pour the venous blood into each branchial heart. The right vena-cava or right branchial vein also receives a vein from ink-gland and gonads. Each branchial heart pumps the blood to the ctenidium of its side by an afferent branchial vein, which runs through the axis of gills and gives off branches. The oxygenated blood from the ctenidum is returned by an efferent branchial vein first to the auricle and then to the ventricle.

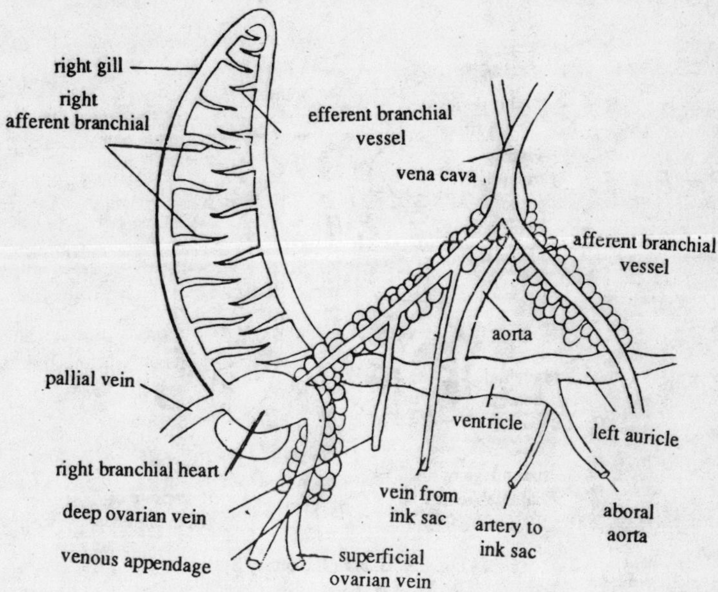

Fig. 6.8 *Sepia.* Circulatory System.

Blood. Blood consists of the plasma containing *haemocyanin,* dissolved in it. Colourless amoebocytes floating in it. It is colourless when deoxygenated and light blue when oxygenated.

Excretory system

The process of excretion is affected by a kidney or renal sac which consists of three thin-walled chambers communicating with one another. Out of them, one is dorsal and median while the other two are ventral and symmetrically situated on either side of the rectum. The two ventral parts of the kidney open to the outside by the renal apertures situated on the renal papillae. Through each ventral chamber passes the corresponding branchial vein of vena-cava. The vein is attached to the dorsal walls of the ventral renal chamber and are covered by glandular epithelium which is excretory in nature. The nitrogenous waste products are extracted by glandular walls and reaches into the renal sac. The dorsal chambers encloses the *pancreatic follicles* covering and opening into the ducts of the digestive gland. They are richly vascular and are said to serve an excretory function. The nitrogenous excretory matter in the cavities of renal sac is in the form of *guanin* which is poured into the mantle cavity.

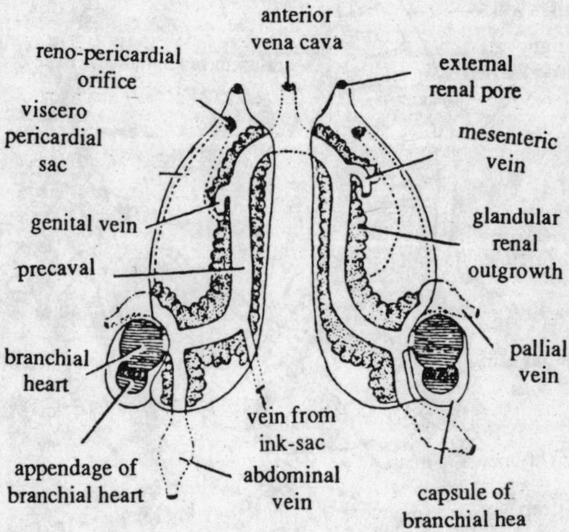

Fig. 6.9. *Sepia.* Excretory System.

Nervous system

The nervous system shows a higher grade of organisation. The nervous system consists of ganglia, connectives and commissures. Following are the ganglia

1. Cerebral ganglia
2. Pedal ganglia
3. Pleuro -visceral ganglia
4. Stellate ganglion or pallial ganglion
5. Gastric ganglion.

1. *Cerebral ganglia.* It is relatively larger is size paired cerebral ganglia are found together into a rounded mass lying in a hollow of the cranial cartilage and covered over anteriorly by a strong fibrous membrane. The brain gives off from the sides a pair of very short, thick *optic nerves* that at once expand into large *optic ganglia* lying in close contact with eyes. A small *olfactory ganglion* lies close to the optic ganglion and sends a fine *olfactory nerve* to the olfactory pits. The cerebral ganglia are connected

with the *superior -buccal ganglia* anteriosly by thin *cerebro-buccal commissures* . The superior buccal ganglia are connected with a pair of closely united *inferior buccal ganglia* by a pair of *circumoesophageal connectives* join the brain with pedal and pleuro-visceral ganglia that lie posterior to oesophagus. A pair of nerves extend from the brain to the statocysts by way of the pedal ganglia.

2. *Pedal ganglia* The pedal ganglia are fused into a mass. They send 10 *branchial nerves* to the arms. Each nerve wxpands into a branchial ganglion at the base of the arm it supplies. The pedal ganglia also give off a pair of *infundibular nerves* to the funnel. The pedal ganglia are connected with the superior buccal ganglia by *bucco-pedal connectives* and with the cerebral ganglia by *cerebro-pedal connectives.*

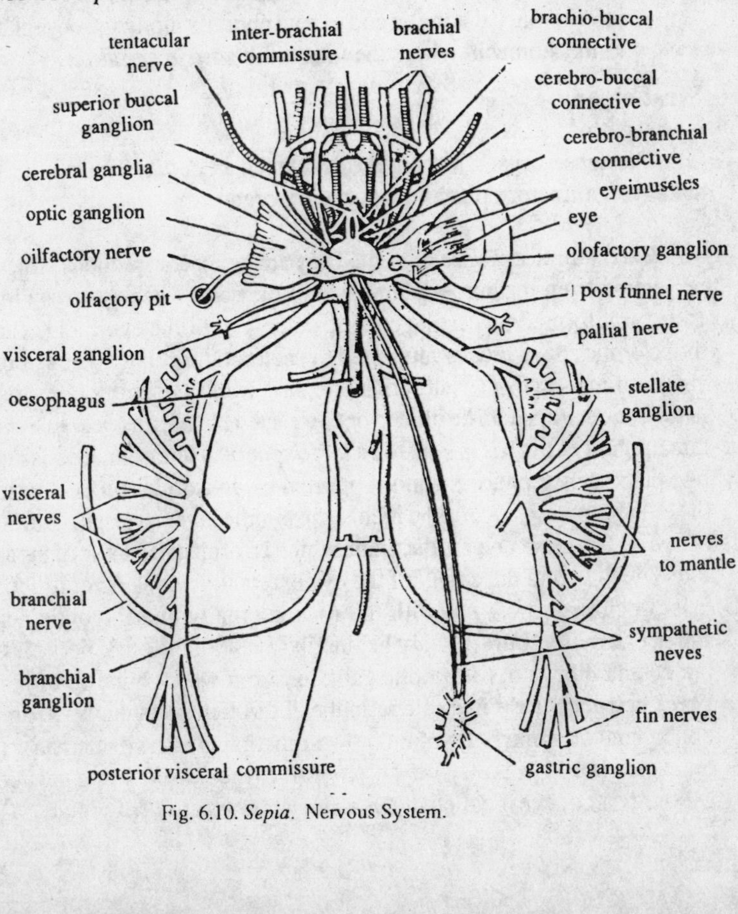

Fig. 6.10. *Sepia*. Nervous System.

3. *Pleuro-visceral ganglia.* These are also united into a single mass being in immediate contact with the pedal behind the oesophagus. The pleurovisceral ganglia give off two *visceral nerves* innervating the various internal organs and forming a visceral loop from which arise a pair of *branchial nerves,* having each a *branchial ganglion* at the base of the ctenidium. The pleuro-visceral ganglia also give off two very stout *pallial nerves* which innervate the mantle.

4. *Pallial ganglion.* Each pallial nerve runs through the neck to the inner surface of the mantle cavity and expand into a large, flat, *pallial* or *satellate ganglion* which is visible in front of the ctenidium when the mantle cavity is exposed.

5. *Gastric ganglion.* A pair of of sympathetic nerves arise from the inferior buccal ganglia and extend aborally along the oesophagus to the stomach, where they meet the *gastric ganglion.*

Sense organs

The sense organs are highly developed in *Sepia*. These are eyes, statocysts, olfactory pits and a gustatory organ.

Eyes. Paired eyes are present. They are large, conspicuous structure projecting from the dorso-lateral side of the head. Each eye is enclosed in an orbit formed by cartilages. The outer wall of the eye-ball is called the sclerotic coat which is supported by seclerotic cartilage. The sclerotic coat continues in front as the contracted iris with an opening called pupil in the centre. Pupil can be dilated or contracted by the muscles. Just in the inner region of the iris is situated a large spherical lens which consists of two plano convex halves aqnd supported by an annular process termed the ciliary process. Choroid which is the middle layer is absent in the eye of *Sepia*. The inner layer is the retina which is formed of a layer of parallel rods which bound the cavity of the eye externally with a layer of retinal cells in close association with the rods internally and the optic nerve fibres externally. Immediately behind the eye-ball the optic nerve is enlarged and dilates to form optic ganglion from which bundles of nerve fibres originate and innervate retinal cell through sclerotic wall. Small optic gland or white body of unknown function is also situated near the optic ganglion. The transparent horny part of the skin covering the upper exposed part of eye is called the *secondary cornea* or *false cornea*. A fold of skin can be drawn like an *eyelid* over the false cornea to some extant. The cavity of eyeball is divided into two unequal parts: a small *anterior chamber*, filles with a watery *aqueous humour*, and a large posterior *chamber* filled with a jelly-like *vitreous humour*.

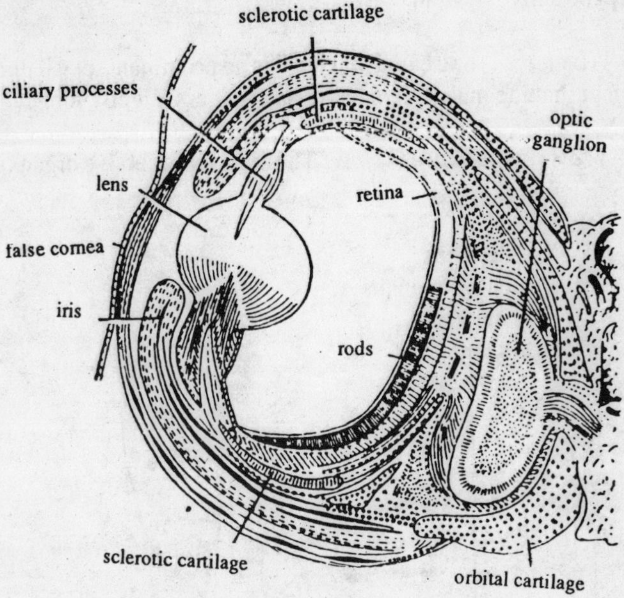

Fig. 6.11. *Sepia*. V.S. of eye

Thus the eye of *Sepia* resembles those of vertebrate eye. However, the vertebrate eye is formed by the outgrowth of brain whereas the cephelopod eye is formed by an ectodermal invagination.

Statocysts. A pair of statocysts, which are organs of equilibrium, lies ventral to the pleuro-viseral ganglian. It remains enclosed in cranial cartilage. A statocyst is a rounded sac, about 3 mm. in diameter. The inner surface is uneven and covered by flat cells, except on posterior side where two spots occur. These are *crists a statica* and *macula statica*. These have large processes at the free ends and the other end produced into fine fibril that forms the statocyst nerve. A large dense *statolith* lies within the statocyst attached to the macula.

Olfactory pits. Small ciliated olfactory pits are situated posterior to the eyes. The sensory cells of the pits are innervated from small olfactory ganglia lying close to the optic ganglia.

Gustatory Organs

A small elevation covered with papillae on the floor of buccal cavity just infront of the odontophore is said to be the organ of taste.

Reproductive System

The *sepia* is unisexual, but there is no prominant sexual dimorphism except that the males are smaller and possess slightly longer arms.

Male reproductive organs : The male reproductive organs comprise

(i) Testis
(ii) Vas deferens
(iii) Vesicula seminalis
(iv) Prostate gland
(v) Needham's Sac

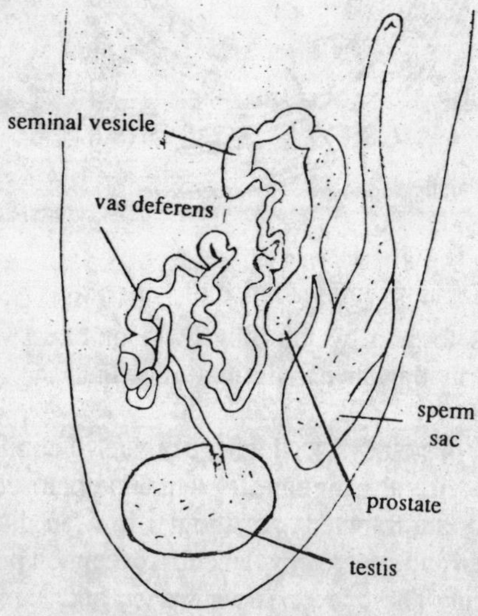

Fig. 6.12. *Sepia.* Male reproductive organs.

(i) *Testis.* The testis is large, oval, yellowish and saccular structure lies near the apex of visceral mass. It is enclosed in a coelomic sac and opens into it by a small apertune on its ventral side. The sperms produced in the testis are passed into the coelomic sac. The coelomic sac leads on the left side into a vas deferens.

(ii) *Vas deferens*. It is a narrow coiled tube, it leads into seminal vesicle.

(iii) *Vesicula seminalis.* It is a long tube. Here the sperms are rolled up into long and narrow bundles enlcosed in ealborate chitinous capsules, called the *spermatophore*. A spermetophore is like an automatic bomb, at one end it has a complex spring-like arrangement which ruptures it wall as soon as transferred to water

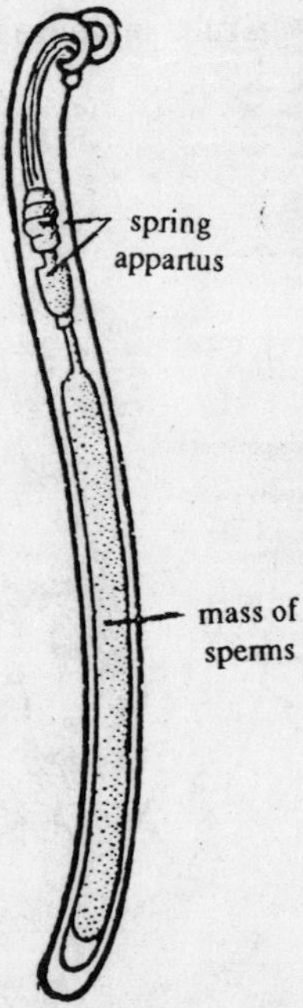

Fig. 6.13. *Sepia* A spermatophore

(iv) *Prostate glands*. The terminal end of vesicula seminalis gives off two blind folds, one of them is called the prostate gland.

(v) *Needham's sac.* The seminal vesicle terminates into a wide reservior, the Needhams sac lying on a papilla or the *penis* to the left of the anus

Female reproductive organs. The female reproductive organs comprise:

(i) Ovary

(ii) Oviduct

(iii) Nidamental gland

(iv) Accessory Nidamental glands

(i) *Ovary.* A large ovoid ovary situated, like the testis in the male , in aboral region of the trunk. It is enclosed in a similar coelomic sac. It has an axial swelling thet bears numerous follicles. Each follicle with a singel ovum supported on a stalk.

(ii) *Oviduct.* A wide thin-walled oviduct arises from the capsule or coelomic sac and runs orally to open into the mantle cavity by *genital aperture.*

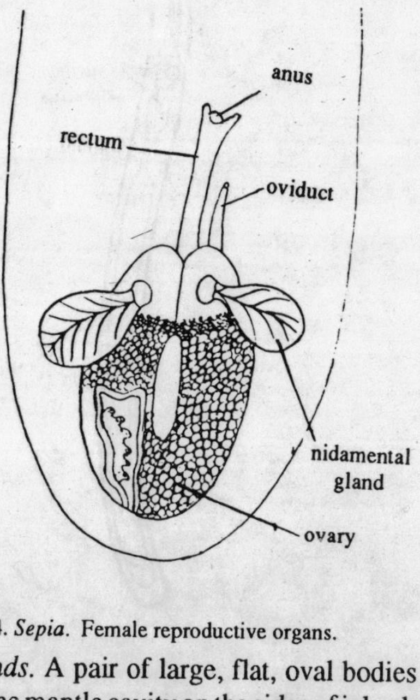

Fig. 6.14. *Sepia.* Female reproductive organs.

(iii) *Nidamental glands.* A pair of large, flat, oval bodies lie on the anterior wall of the mantle cavity on the sides of ink-gland . These glands secrete a viscid material by means of which the eggs when deposits adhere together is masses.

(iv) *Accessory Nidamental glands.* A glandular mas of organe colour known as accessory nidamental glands lie at the sides and around the anterior sides of the nidamental glands. They open into the mantle cavity by a number of minute pores. Their function is not clear.

Copulation, fertilization and laying

In *Sepia* the 5th left arm of the male is *hectocotylized* i.e. modified for cooulation by reduction of some of the basal rows of suckers. The male inserts it in its own mantle cavity to extract the spermatophores. Later on, this arm is inserted into the mantle cavity of the female and the spermatophores are deposited in *bursa copulatrix*, which is a modified part of the funnel. The speremetophore are burst by the spring apparatus in them, and release the sperms after copulation. When ripe ova are passed out side one by one from the ovary through the coelomic funnel and fertilized. The viscid secretion of Nidamental gland covers the eggs and cement them into a mass. These masses of eggs are attached by stalks to the twig of marine plants or weed or some other object.

The development is direct without larval stage.

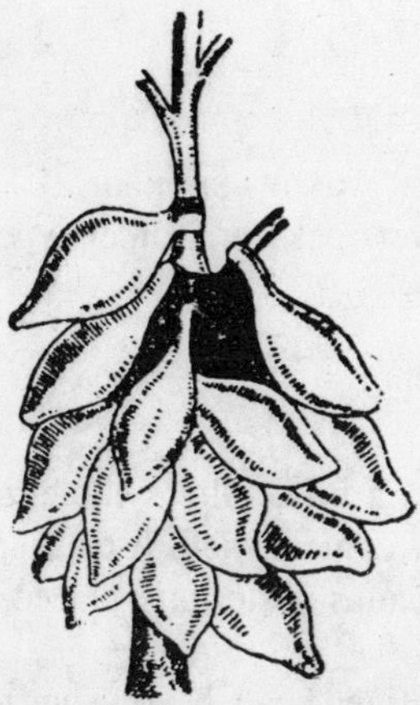

Fig. 6.15. *Sepia*. A egg mass.

7

ARCHIMOLLUSCA

It appears almost impossible to acquire the size knowledge about the primitive form of mollusc, yet such a structure had been hypothetically reconstructed by careful observation on the account of morphological and anatomical researches. The discreption of the ancestor mollusc is based on the work of several workers. But largely it represent the views of *Yonge*. No living mollusc, even *Neoplina,* is like this ideal hypothetical ancestor.

Habits and habitat.

The ancestral mollusc was a inhibitant of Pre-Cambrian oceans where it moved about over rocks and other types of hard bottom in shallow water.

External features.

The archimollusc had a bilaterally symmetrical body which was dorsally covered by a low conical shell. It measured 2.5cm. in length. Mouth was anteriorly and anus posteriorly placed.

Head. It had a well defined head, bearing one pair of tentacles, with one pair of *eyes or ocelli* at their bases. The mouth occupied the ventral position.

Foot. It had a ventral creeping foot with powerful and well-developed musculaturc *sole*. Around the base of foot runs an *epipodial fringe*, beset with delicate tactile tentacles.

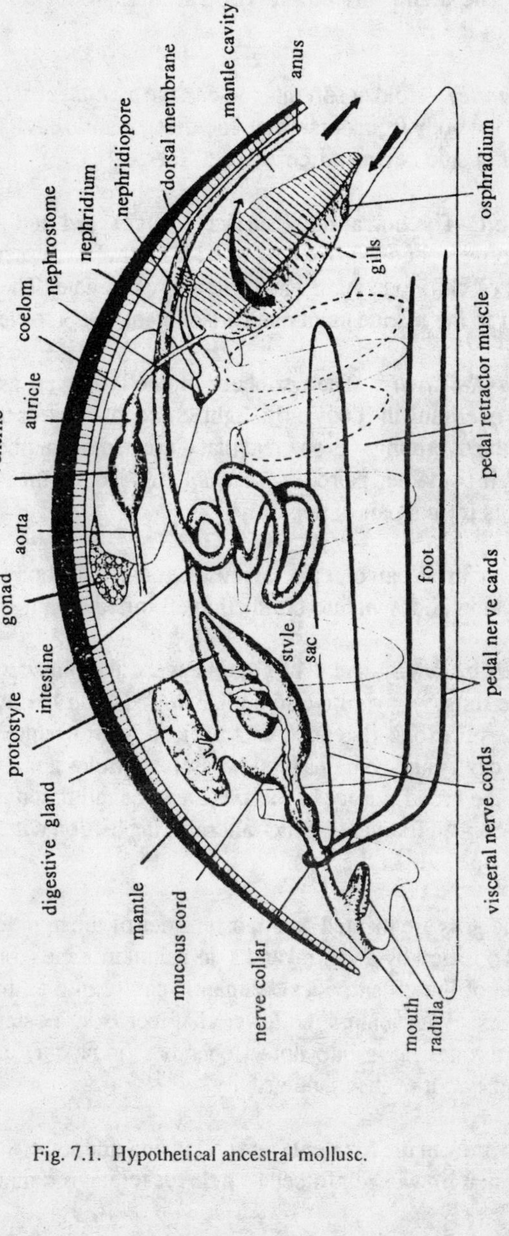

Fig. 7.1. Hypothetical ancestral mollusc.

Visceral mass. The dorsal and posterior part of the raised like a hump. The hump formed the visceral mass which included the vital organs of the body.

Mantle. A fold of soft integument, the *mantle or pallium* hang down around the body from dorsal side enclosing mantle cavity and communicating out side between free edges of foot.

Shell. The dorsal body surface was covered and protected by an oval, convex or conical external shell. The shell was secreted by mantle consist of chitinous matrix and calcium carbonate. The shell provided support to the mantle cavity and attachment to dorso-ventral muscles.

Mantle Cavity. The periphery of shell, as well as its underlying mantle overhung the body only slightly, except towards the posterior end where the overhung was so great that it created a chamber called mantle Cavity. It contained paired *ctenidia and ospharidia*, the median *anus* and openings from a pair of nephridia.

Muscles. A pair of retractor muscles enable the body particularly the foot to be withdrawn into the shell to a limited extant.

Respiratory System. The gills of ancestral mollusc were very probably like those of primitive living gastropods and bivalves. Such a gill consists of a long flatenned axis projecting from anterior wall of the mantle cavity and contains blood vessels, muscles aqnd nerves. To each side of the broad surface of the axis are attached flatenned triangular or wedge-shaped filaments, which alternate in position with those filaments on opposite side of the axis.

The gills are located on opposite sides of the mantle cavity and are held in position by a ventral and a dorsal membranes. Functionally, the position of the gills divides the mantle cavity into an upper and lower chambers. Water enters the lower chamber from posterior, passes upwards through gills into dorsal chamber (exhalant) and then moves posteriorly out of the mantle cavity.

To prevent the filament from collapsing under the flow of water, the frontal margin was reinforced by a tiny, skeletal rod made up of chiton.

Two blood vessels run through the gill-axis. The afferent vessel, which carried blood into the gill, and the *efferent vessel*, which drains the gill.

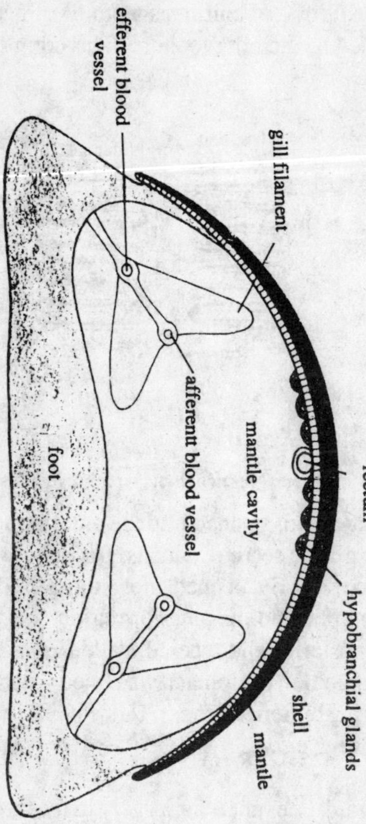

Fig. 7.2. Hypothetical mollusc in T.S.

Body Wall. In the ancestral mollusc the mantle epidermis and the epidermis of reminder of exposed body parts, including foot, were covered by cilia and contained mucous gland cells. Mucous helped in locomotion as lubricated the substratum.

Digestive System. The anterior mouth opened into a chitin lined buccal cavity. The posterior wall of the buccal cavity was evaginated to form the pocket-like redula-sac. Redula sac contained on its floor the *redula*. The redula apparatus made up of an elongated cartileginous base, the *odontophore*. Over the odontophore and around its anterior was stratched a membraneous belt, the redula proper, which possessed a number of longitudinal rows of chitinous teeth. Protractor and retractor muscles were attached to the odontophore. The odontophore could be thrust out of mouth.

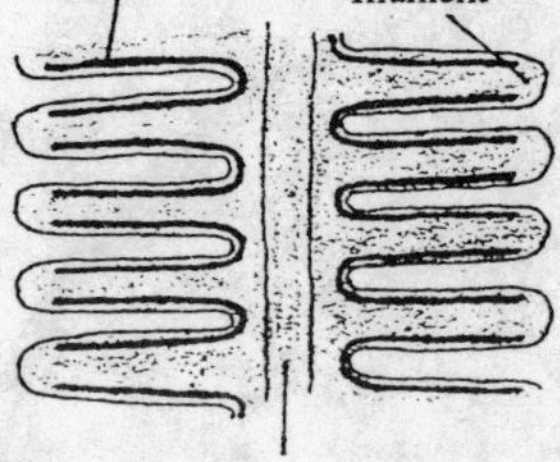

Fig. 7.3. Frontal Section through primitive gill

A pair of salivary glands opened into the anterior dorsal wall of the buccal cavity. These glands secrete mucous. Buccal cavity followed by oesophegus which porteriorly opened into stomach. The stomach of archimollusc was complicated. Its anterior region was broad and hemispherical where as posterior end tapered, called *style sac*, led into the intestine. The anterior region of stomach was lined with chitin except for a ciliated, ridged sorting region and the entrance point for two ducts from a pair of lateral *digestive glands*.

The long coiled intestine functions largely in the formation of faecal pellets. The anus opened mid-ventrally at the posterior margin of the mantle cavity and the wastes were swept away by the exhalant current.

In all probability, the ancestral mollusc was a microphagous herbivore that fed on fine algae growing on rocks in shallow water.

Coelom. The coelomic cavity was located in the mid-dorsal region of the body. The Cavity surrounded the heart dorsally and a portion of intestine ventrally. Thus the coelom represented both a pericardial and a perivisceral coelom.

Circulatory System. The heart comprised a pair of posterior *auricles* and a single muscular anterior *ventricle*. The auricles received oxygenated blood from the gills and than pass it into the ventricle. Ventricle pumped blood through a single aorta. the aorta branched and rebranched into smaller blood vessels that delivered blood into *sinuses*, where the tissues directly bathed directly. Thus the circulatory system was open type. From sinuses blood travelled to the gills and back to the auricles. Plasma contained colourless amoebocytes as well as the respiratory pigment called *haemocyanin*.

Excretory system. The excretory organs of the archimolluse probably consisted of a pair of tubular *metanephridia*, commonly called kidneys in living species. Each nephridium opened at one end into the pericardium through a *nephrostome*, and at the other into the mantle cavity through a *nephridiopore* that was located laterally to the intestine. The pericardial coelom received wastes from two sources. The heart wall delivers a filterate from blood, and glands in the pericardium secrete waste into the coelom. The pericardial fluid then passed through the nephrostome into the nephridium. The nephridium discharged into the mantle cavity through the nephridiopore.

Nervous system. Nervous system was composed of a nerve ring around the oesophagus which may or may not had ganglia. From this nerve ring arose two pairs of nerve trunks running longitu-dinally through the body with ganglionic cells distributed along the entire length. The ventral pair represented the pedal cords and dorsal pair to the visceral cords innervating the mantle and visceral organs.

Sense organs. Sense orogans consisting of paired tentacles, eyespots, a pair of statocysts in front and two osphradia. The osphradia were the patches of sensory epithelium, located at the bases of gills to test the quality of water.

8

FOOT IN MOLLUSCA

The ventral side of the body, in Mollusca, is characterised by the pronounced development of its musculature, which enables the animal to creep. A fleshy foot, provided with a flat sole,is suited for such a locomotion. The foot is quite distinct from rest of the body, especially from the head.

The flat form of the foot with a sole for creeping must be considered the primitive form, such a foot may be found in Chitons among the class-Amphineura, but it is variously modified among various classes of Phylum Mollusca to perform diverse types of functions i.e. swimming, attachment, digging and so on. The foot and its appandages get their nerve-supply from the pedal ganglia or the pedal nerve cord.

The foot may be distinguished into various parts as follows:

(i) *Anterior Propodium.* It is the anterior distinctly marked portion from rest of foot.

(ii) *Metapodium.* It is the posterior part of foot and carries operculum.

(iii) *M*esopodium. It is the central portion between the first two.

Apart from this some times there occurs expanded lobe-like margins of the ventral sides of foot called as *parapodium*, which can be folded and united over the head and sometimes there are paired, prominant ridges arising from the side of the base of foot and called *epipodium*, which are provided with sensory tentacles and eye-sports. In caphalopods the foot becomes modified into the tentacles, suckess and siphons.

Variations of foot in different groups of mollusca is as follow :

1. Foot in Monoplacophora.

In *Neoplina* the ventral surface bears a prominant foot in the centre. The foot shows a primitive condition. It is broad, circular, with a flat sole on which the animal creeps by muscular waves.

2. Foot in Amphineura

Class Amphineura is divided in two subclasses. Placophora and Aplacophora.

(i) *Placophora.* it possesses most primitive of foot which is large muscular organ with a flat, creeping surface extending along the whole length of body on its ventral surface. The foot is not differentiated into different parts. The example is *Chiton.* Besides performing locomotary function of creeping, it also serves as an organ of adhesion to the surface of rocks, when at rest.

(ii) *Aplacophora.* The foot is vestigial and is represeted by a aciliary ridge running along the base of narrow, median, ventral groove, extending between the mouth and cloacal aperture., In some forms like *Chaetoderma* both, the foot and ventral groove are absent.

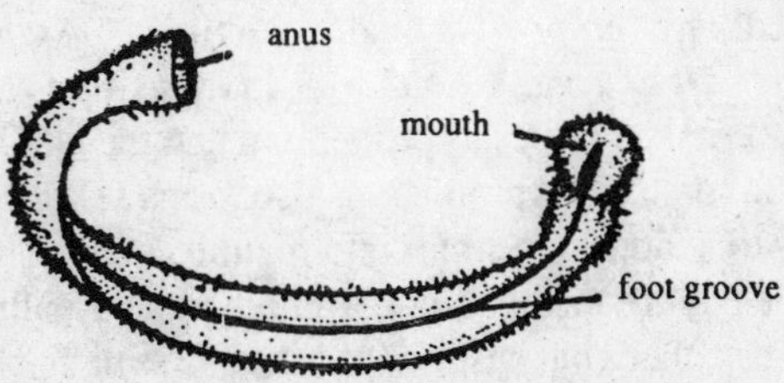

Fig. 8.1. Foot in *Chaetoderma*

3. Foot in Scaphopoda.

Foot is almost cylindrical or conical as in *Dentalium* and projects downward into the mantle cavity through the lower aperture of which it can be protruded out to serve as the digging organ. Engorged with blood, it serves as an anchor. The free end of the foot is conical and trilobed carrying on each lateral side a wing-like fold or *pleat* which is comparable to the *epipodium*. A groove runs along the middle line of foot.

In *Siphonodentalium* both, the groove and lateral folds are absent and the anterior end of foot bears a contractile disc with papillated margins, while in *Pulsellum*, a filiform tenacle springs from the middle of the disc.

4. Foot in Gastropoda.

The foot of gastropods contains scattered unicellular as well as aggregated multicellular glands which secrete mucous. These glands are called *pedal glands*. The pedal glands include:

(i) *Labial gland* in a groove at the anterior end,

(ii) *Suprapedal gland* between head and anterior end of foot (iii) *Sole gland* opening mid-ventrally in the middle of foot and (iv) *posterior glands* at the hind end of foot.,

(A) *Foot in subclass Prosobranchia.* The foot in different prosobranchiates may becomes variously modified in relation to different conditions of life.

In several pectinibranchiates the anterior portion of foot is only well differentiated into *propodium* and sometimes *metapodium* equally developed, *e.g. Natica and Xenophorus*. In these forms when the animal is in motion it first advances its propodium and then folds the rest of the foot after it. In some forms propodium also forms a siphon on the left side which helps to conduct the water into mantle cavity. The operculum is always carried by the metapodium. Epipodium is usually absent among the pectinibranchiates but vestiges of it can be seen in the form of a pair of anterior pedal tentacles in *Vermitus*, in the form of two nuchal lobes in *Ampullaria*.

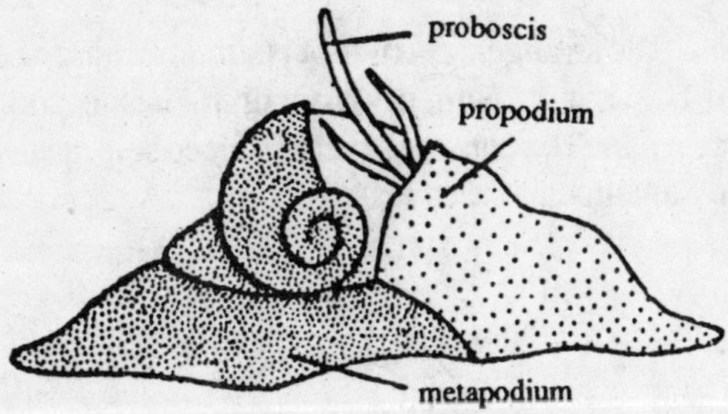

Fig. 8.2. Foot in *Natica*

In aspidobranchians a well developed epipodium is always present. A good example is met in the case of *Haliotis* in which epipodium surrounds the base of foot in the form of ridges with digited appendages and long sensory contractile tentacular processes. These may be provided with sensory organs.

Broadly speaking, most of the prosobranchiates creep or attach by means of flat sole but in several pelagic pectinibranchiates foot has become especially modified in accorodance with swimming mode of life and propodium is changed into a narrow vertical fin and the development of fin can be traced almost step by step starting from *Oxygyus* proceeding through *Atlanta* and *Pterotrachea.*

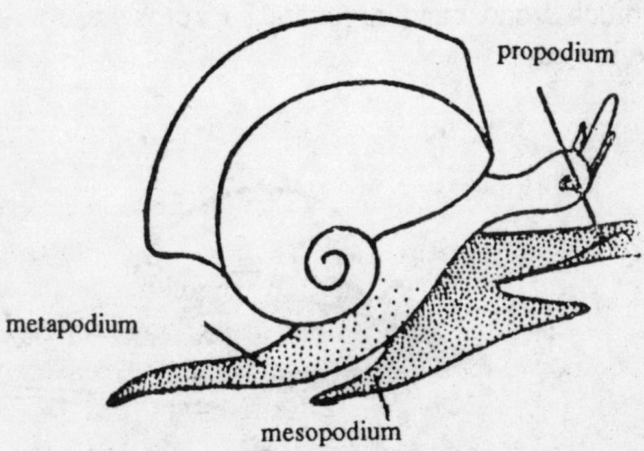

Fig. 8.3. Foot in *Oxygyrus.*

(a) *Oxygyus.* Foot consist of propodium which anteriorly possesses a fin-like out growth and a distinct metapodium directed backward like a tail and bearing operculum. The sole is mainly used as sucker.

(b) *Atlanta.* The arrangement of foot is similar to that of *Oxygyus*, but the fin-like out growth of propodium has become its most important part. The comparatively reduced sole or sucker appearing only an appendege to it.

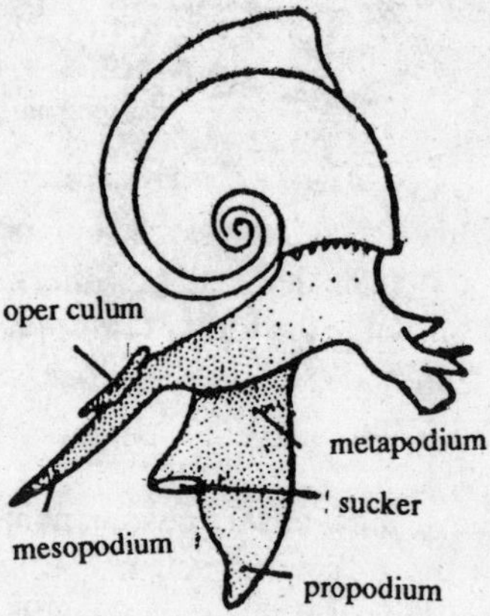

Fig. 8.4. Foot in *Atlanta*.

(c) *Carinaria.* Both, the foot and general external appearence of entire animal, are much changed. The metapodium which here has no operculum appears as a posterior prolongation of body. The fin is much broader and larger and sucker is present just along 1/3 of its free edge.

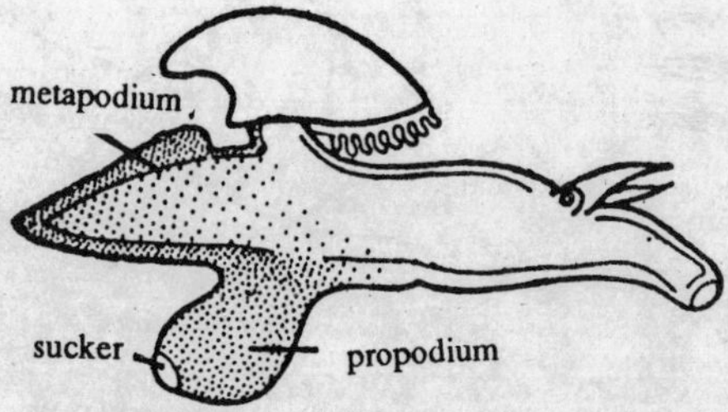

Fig. 8.5. Foot in *Carinaria*

(d) *Pterotrachea*. The sucker pesists only in male, while the metapodium loses the operculum and becomes reduced to a short, filament like tail.

(B) *Foot in Sub-Class Opisthobranchiata*. While there is no division of foot into various part and among the epipodia are lacking but parapodia are highly developed. These parapodia are lateral lobes of fold-like extensions of the edges of the sole of foot and they are often bent back over the shell so that their edges sometimes touch each other. Thus the shell is completely covered by them. These parapodia may even fuse along their upturn edges at the posterior end. Such a fusion is remarkable in *Aplysia and Notarchus*.

Among plegic opisthobranches the parapodia develop into paired fins and foot may become confined the anterior end of the body and two lateral fins are used in locomotion.

In nudibranches the foot is no more exist as a differentiated organ but the body of animal is compressed in the shape of a leaf with sharp dorsal and ventral edges to enable the animal to move forward *e.g. Eolis*.

(C) Foot in Subclass Pulmonata. The foot is always undivided and provided with a large, flat sole for creeping. In few forms, however, it is divided into two portions by a temporary or permanant groove.

5. Foot in Pelecypoda.

Foot, like that of majority of mollusc, is a locomotory organ present as a muscular projection from the ventral surface. Usually in this class the foot is painted axe-shaped with sharp edge directed downward and outward. In Protobranchiate for examples *Nucula Yoldia* the foot has a flat, creeping surface and is considered the most primitive foot in the class.

The foot of lamellibranchiates varies according to the mode of locomotion and according to development of *byssus*. It is secreted by the byssus gland and consists of large number of threads vary in thickness. With the help of these fibres the animal anchors to the foreign bodies.

The complete byssus apparatus consists of:

(a) A Cavity in the foot into which the byssus gland opens.

(b) A duct connecting the cavity to exterior.

(c) A groove which runs from the opening of the duct along the ventral edge of foot to its anterior end.

(d) A crescent shaped or cup-like widening of the groove at its anterior end where the threads are present.

The relation existing between the development of foot and that of byssus apparatus may be given as follow.

(i) The foot is in its primitive form with a flat sole and no groove and with a simple invagination but without byssus *e.g. Solemya.*

(ii) With the same foot small lamella arises from the base of the simple invagination and byssus is very slightly developed. e.g. *Nucula.*

(iii) Invagination becomes differentiated into a cavity and the byssus and its gland are highly developed. *e.g. Mytilus.* In *Arca* and *Mytilus* the old byssus may be shed and a new one Formed.

(iv) Many lamellibranches in the adult stege have neither byssus nor byssus gland but the cavity and the duct and even the retracter may be retained. Byssus apparatus may be found into close related forms without the byssus itself and sometimes with it. In the former case the foot is generally more developed and serves for locomotion e.g.*Trigonia*.

(v) The wedge-like or blade-like foot often becomes more strongly developed as a fleshy and extensile organ every trace of byssus and its apparatus are absent. e.g. *Unio, Cardium, Tellina* etc. The anterior end of the foot forms a long slendrical tentacle in *Poromya*, with a swollen free extremity in *Axinus*.

6. Foot in Cephalopoda

The foot in cephalopoda is represented by arms and siphons. The arms are regarded as the lateral processes of the molluscan foot which have reaches beyond the head and united infront and the head is encircled by the foot and the mou th comes to lie in the centre of this. This circle of arms is derivative of the foot which is proved by its anatomical and ontogenie evidences which are as follows:

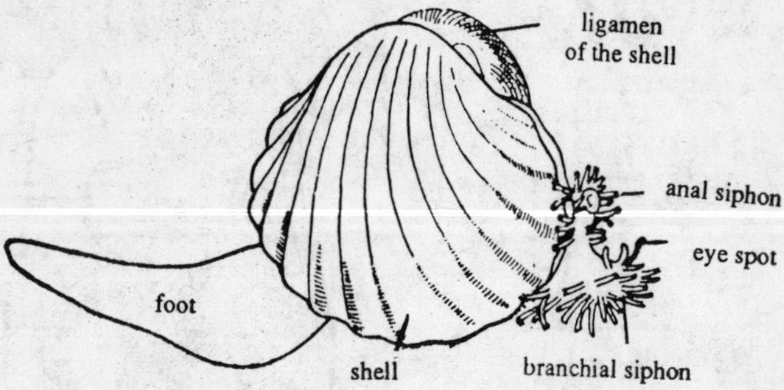

Fig. 8.6. Foot in *Cardium*

(i) The arms or tentacles are innervated by the brachial ganglion which is anterior origin of pedal ganglion and it lies under the oesophagus.

(ii) The arms do not occupied in the embryo their definitive possession from the mouth but arise on the ventral surface behind the mouth and between the mouth and anus in a row found each side. These two rows shifts to form a circle of arms in the adult condition.

(A) *Foot in subclass Dibranchiata.* In Dibranchiata 8 or 10 such arms are present provided with two rows of suckess. In *sepia* more than two rows on ventral side. These suckers are stalked in Decapoda and sessile in Octopoda.

The octopoda has eight arms connected by means of membrane at their bases. In decapods these are ten arms, eight of which correspond to the eight armed octopods but they are not connected by the membrane at their bases. The remaing two arms are longer with swollen ends and their suckers being confined to their free ends. Each of these two arms is situated between 3rd and 4th Octopodus arm on either side. They lie retracted into special cavities of head and thrown out of their cavities with great force, while catching the food. One, rarely both, of arms is modified to assist in copulation and called as hectocotylized arm.

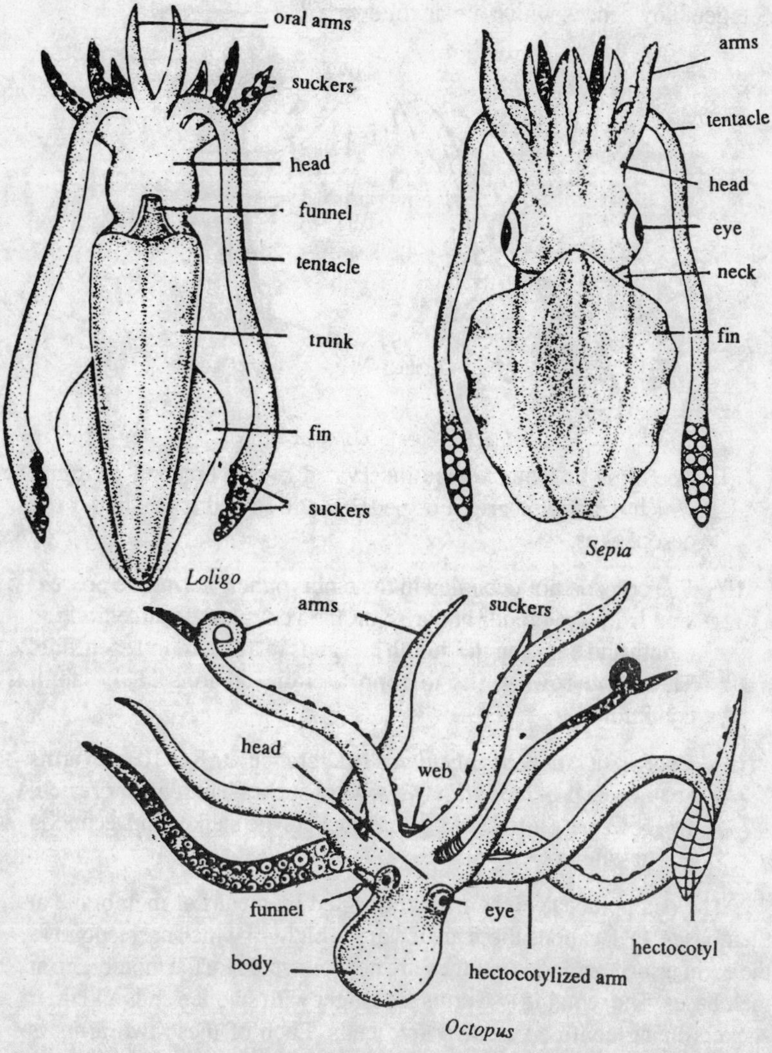

Fig. 8.7. Foot in *Caphalopoda*

(B) *Foot in subclass Tetrabranchiata.* The head in *Nautilus* bears several tentacles arranged in circle towards the mouth about 90 in females and 60 in males. These tentacles are present on special lobes which may be taken to correspond to the arms of dibranchiates while tentacles to the suckers. These tentacles are adhesive and prehensile and are retractile within special tentacular sheaths. In the female there are three tentacular lobes in immediate contact with buccal aperture, these are right, left and ventral-anterior lobes. In male the ventral-anterior lobe is absent instead it bears *Owen's organ*, which is supposed to be olfactory in nature.

In addition to foot proper the cephalopods possess epipodia which are highly modified to form the funnel or siphon. In *Nautilus* the funnel consists of two symmetrical lateral lobes which are enclosed towards each other and overlap sometimes. But in dibranchiate these lobes fuse with each other to form a complete tube projecting beyond mantle cavity. Usually the anterior lobe of funnel is provided with a velum *e.g. Nautilus* and several decapods but absent in octopods.

9

RESPIRATION IN MOLLUSCA

The requirement of oxygen presents little difficulty to small organisms in which diffusion is adequate to carry requirement of oxygen to the cells. It is among the molluscan animals with their relatively large size and covered body that we first meet distinct specialized respiratory structures. Primitively these structures were probably simple projections of the epithelium of mantle but in most present molluscs these projections called as *ctenidia*. The ctenidia are greatly modified in the direction of increase in surface area and greater vascularity. The second important development among respiratory organs found in connection with the ability of certain gastropods to live permanantly on ground. In pulmonates ctenidia has disappeared and the walls of mantle cavity have folded to form a sac-like structure which is highly vascular. So in molluscs there are three types of respiratory organs.

1. Ctenidia
2. Pulmonary sac
3. Skin

1. Ctenidia.

Ctenidia are the organs of aquatic resspiration and is mostly present in aquatic molluscs. The pulmonary chamber or sac is the organ of aerial respiration and is present in terrestrial molluscs. But in certain cases like *Pila, Planorbis* etc. both the methods of respiration are employed. Some animals may be regarded as in process of change from an aqueous to tessestorial life and as we see in *Siphonaria*, the ctenidia are reduced to a great extent since the animal comes out of water on the average 22 hours out of 24 hours.

Structure of Ctenidium. The ctenidium has a *central axis* one or two rows of *lamellae or filaments*. The entire gill is surrounded by cilia. The ctenidium receives venous blood through an affarent branchial vein and after oxygenation in filaments it is send back through an efferent branchial vein.

Number of gills. The number of gills vary from 1-80 in different classes of Mollusca.

Types of gills. In molluscs following types of ctenidia are seen:

1. Monopectinate
2. Bipectinate
3. Plicate
4. Holobranchs
5. Mesobranchs
6. Secondary gills.

1. *Monopectinate.* The filaments or lamellae are arranged in a single row on one side of the axis.
2. *Bipectinate.* The gill has two rows of gill filaments on the sides of central axis.
3. *Plicate gills.* These are transverse folds of integument.
4. *Holobranchs.* The ctenidia extend all over the body.
5. *Merobranchs.* When the ctenidia are restricted to a particular area of the body.
6. *Secondary gills.* The true gills are absent. But certain other structures funchion as gills. They are the following.

(i) *Anal gills.* Gills present around the anus.

(ii) *Cerata.* These are slender vascular bodies located on the dorssal surface.

(iii) *Pallial gills.* Gills arranged in single row around the foot.

In Monoplacophora

In class Monoplacophora, 5 pairs of gills are present. These are unipectinate, composed of finger-like lamellae. The gills are present in pallial groove. Vestigial gills are supoosed to be present on the opposite side of the ctenitial axis. The gills are probably not true gills, more likely they represent prectenidial structure *e.g. Neoplina.*

In Amphineura. In order Placophora (Chitons) the ctenidia are arranged in two rows. The number vary from 6-80. According to species when their number is few, the ctenidia are mostly confined to the poosterior end. It is not easy to decide whether the multiple form modification found in Chiton is a simple or evolved one.

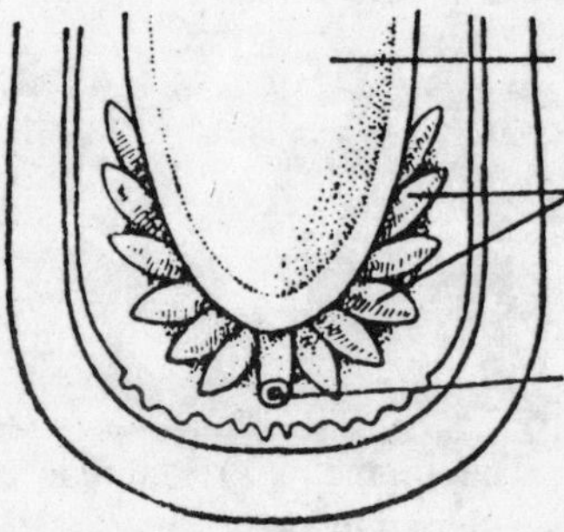

Fig. 9.1 *Chiton.* Showing gills

In the animals of order Aplacophora the foot is absent and the mantle Cavity is greatly reduced in the form of a cloacal Cavity at the posterior. In *Chaetoderma* there are two small, feather-like gills symmetrically placed on either side of the anus. In *Neomenia* the gills are further reduced and consist of single bunch of filaments lying below the anus. Such gills restricted to a limited space at the posterior end of the body are called *Merobranchial type.*

In Gastropoda

The gills show a good number of variations started from bipectinate paired structure to monopectinate single structure due to the assymmetry and in some cases certain secondary structures or adaptive gills are developed which subserve the function of respiration. The gills are shifted in front of the body within the mantle cavity as a result of torsion.

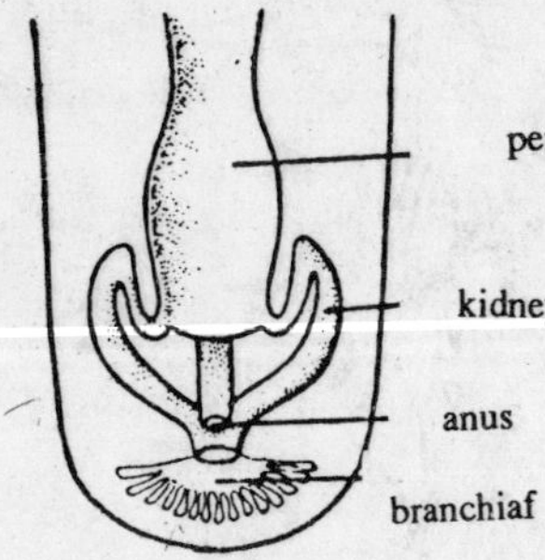

Fig. 9.2 *Neomenia,* Showing gills

In Prosobranch symmetrically paired gills occur only in *Haliotis and Fissurella.* In all others the asymmetry of the body have caused one of the branchi to become reduced and consequently it is only one branchi usually left in vast majority of prosobranchs. Even in *Haliotis* the right branchi is rather smaller than the left. In Aspidobranchs (*Trochus, Turbinus, Acmaea* etc.) the gill is bipectinate but in Pectinibranchs like *Triton, Pila* etc. the gill is monopectinate.

In opisthobranchs a true ctenidium only occurs in the *Aplysia.* It lies on the right and usually more or less external being partly covered sometimes quite shallow and sometimes by a fold of membrane. The Nudibranchs (*Doris, Aeolis)* have altogoether lost the true ctenidium with mantle cavity and respire by various secondary structures such as anal gills (*Doris*) and Cerata (*Aeolis*).

In Patella (Prosobranch) bears a series of adaptive gills occurs in a row on each lateral side in the pallial groove. Such gills are called pallial gills. They may be found together with the ctenidium in *Pneumoderma* (Ospisthobranch).

In Pelecypoda.

The ctenidia are two ~~in n~~umber right and left each consisting of a horizontal axis bearing two rows of filaments. They lie one on each side of foot. They lined a chamber known as branchial chamber. Leading into

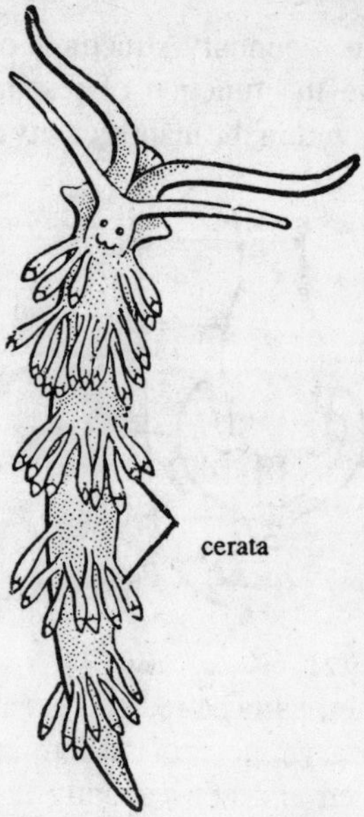

Fig. 9.3. *Aeolis*. Showing cerata.

this chamber and behind it are two siphons, one of which conducts water to the branchial chamber called *inhalant siphon* and other carries the water out side is called *exhalant siphon*. In many cases like *Mya* both the siphons are exceedingly long, sometimes longer than the whole shell. In

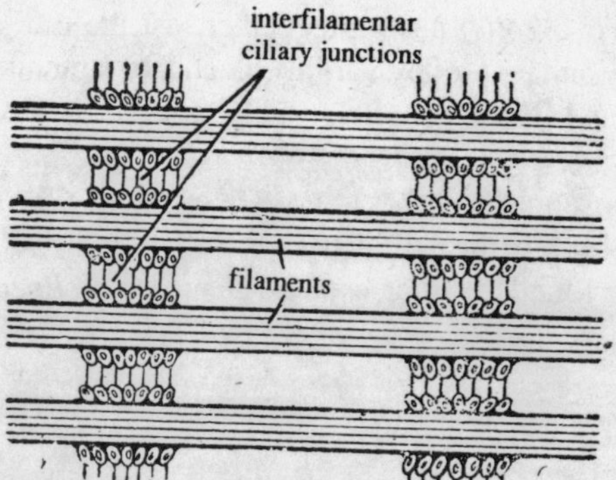

Fig. 9.4. *Mytilus*. Gill filaments magnified.

some cases the two siphons are free throughout their entire length while in ot hers they become fused. In other genera which do not burrow like *Pecten, Mytilus* etc. the siphons are either reduced or altogether absent.

In Protobranchiata *(Nucula)* the gills are relatively smaller, situated at the back of the mantle cavity, and their filaments relatively few, form small, compressed and horizontal disposed, triangular leaflets, free from one another. The leaflets are arranged in two oppoosite rows in a gill. The arrangement divides the mantle cavity into a large inhalant chamber below each gill and a small exhalant chamber above.

In Filibranchia, the gill-filaments become elongated and thread-like. In *Arca* each gill filament is bent upon itself to from an elongated 'V'. Thus each gill forms a 'W' in section and consists of two 'V' shaped demibranchs. Each demibranch is two lemellae thick made up of a descending limb and an ascending limb. In *Mytilus* the gills are strengthened by the development of delicate non-vascular bar or inter-lamellar junctions between the two limbs of each filament. Thus the single ctenidium of *Nucula* is replaced by two plate-like laminae in *Mytilus*. Each lamina is made up of an outer and an inner lamina.

In Psendolamellibranchia (*Pecten, Ostrea*), the gills have a greater cohesion than in filibranchs and the reflected distal ends of filaments become fused laterally with the mantle.

In Eulaniellibrancia (*Unio, Anodonta)* the ciliary interfilamentar junctions are replaced by vascular cross connections with narrow opening between them called *ostia*. The interlamellar junctions also become large and vascular, so that the inter-lamellar space is partitioned into vertical water tubes. The incomming water enters the water tubes through these ostia.

In Septobranchia (*Poromya, Cuspidaria* etc.) the gills are represented by a muscular septum which divides the mantle cavity into the dorsal and ventral parts. This septum is perforated at certain places, which allows the communication between dorsal and ventral chambers.

In Cephalopoda

The ctenidia are situated in the pallial cavity on either side of the visceral mass. In Dibranchia there are only one pair of gills. A gill is

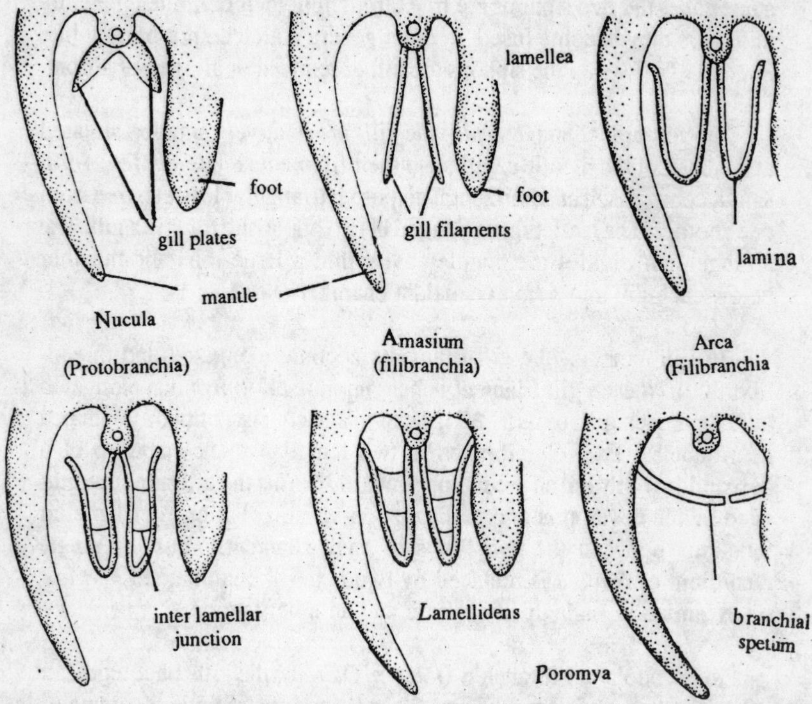

Fig. 9.5. T./S. of pelecypods showing arrangements of gills

composed of lamellae whose number varies in different forms. Each lamellae is thrown into transverse folds which are inturn folded so that the respiratory surface is largely increased. The gills are attached dorsally to the mantle (*Sepia, Loligo and Octopus*). In Tetrabranchia there are two pairs of gills which are free throughout their existence (*Nautilus*).

Histology of gill-filament. The gill-filaments are hollow structures and each filament has two sides. One side facing towards the mantle cavity and other towards the inter-lamellar cavity. The gill filaments are provided with Chitinouss rods to give a sort of rigidity to the gill filaments. They are covered with ciliated epithelium. The epithelial cells are larger on the side facing mantle cavity and secrete mucous. The presence and arrangement of cilia is remarkable, on the sides are *lateral cilia*, on the outer ridge-like faces of filaments are *frontal cilia* and those lying in between are *latero-frontal cilia*.

2. Pulmonary Sac or Cavity.

It is an organ of aerial respiration and is found in land molluscs, mostly in pulmonates and some amphibious prosobranchia. When we use term 'lung'it must be remembered that this organ in mollusc does not correspond morphologically with spongy, cellular lungs of vertebrates. It simply performs the same function. The lung in mollusc is a cavity or pouch lived by blood vessels which are disposed over its surface in various ways forming network.

In adaptation to terrestrial mode of life the pulmonary respiration resulted. Therefore it is found in vary different groups of Gastropoda, but different stages of evolution are best studied in Streptoneura. In this group certain aquatic forms, though they possessed Ctenidia, required the habit of living for short or longer time out of water. Consequently certain modifications have taken place in the internal structure of mantle. So that in certain forms such as *Ampullaria and Pila*, both Ctenidia and pulmonary sac are present and the mantle cavity is divided into right pulmonary chamber and left branchial chamber by an incomplete septum. In others the ctenidia has totally absent and the pulmonary chamber is fully developed (*Gadinia*).

3. Outer Skin.

Respiration by moist-skin is the simplest method of respiration in molluscs. In certain aquatic (Scaphopoda and terrestrial Opisthobranchiata), terrestrial (*Vaginula*) and parasitic (*Ontodax and Entococha*) neither ctenidia nor localized respiratory organs are present. Respiration occurs through entire body surface or soft-skin of the body. In *Oncidium* respiration is, however, carried by skin as well as by dorsal papillae present on skin. Skin function as respiratory organ even when ctenidia, as well as pulmonary cavity is present.

10

TORSION AND DETORSION AMONG GASTROPODS

In all the gastropods with coiled shells the mantle cavity is anterior, the opening directedforward and the coiling of visceral hump is directed posterior. But in the development of these forms from the larva the mantle cavity first makes its appearence behind the visceral hump and at a particular stage the visceral hump rotates in a counter clockwise direction through an angle of 180^0 on the rest of body. This is known as *torsion*. Thus torsion may be defined as the rotation of the visceral organs to an angle of 180^0 in the anticlokwise direction. It occurs in the larva of gastropods and converts the symmetrical larva into an asymmentricl adult.

Process of Torsion

It occurs in two stages:

1. Ventral flexure
2. Lateral torsion

1. *Ventral flexure*. It takes place in the anterio-posterior to the sagittal plane about the transverse axis situated at right angle to the main anterio-posterior axis of the animal. It results in the looping of alimentary canal. As a consequence of this flexure the visceral mass, the shell which was orginally saucer shaped becomes more or less pointed cone and during this process the visceral sac finally assume spiral coiling

2. *Lateral torsion.* The ventral flexure is soon flollowed by lateral torsion and dorsal shell becomes ventral in position. This is brough about by unequal growth on two sides. Generally the organs present on the left side grow normally and of the right side has retarded development. The organs from left shift towards right. The mantle cavity and pallial complex are shifted towards the right side and becomes anterior in position

Effects of Torsion

Present gastropods are asymmetrical showing twisting of nervous system and asymmentry of internal organs. The peculiarities of organisation of gastropods due to torsion were first realized by *Spengel* (1881). The effects of torsion are not equally evident in all gastropods. Following changes are seen after torsion:

1. Before torsion the ctenidia, anus,renal aperture point backward and the auricles lie behind the ventricle. After torsion ctenidia project forward, anus lies an right side and auricles lie anterior to the ventricle. The original posterior face of the visceral sac becomes the anterior face.
2. Nervous system becomes twisted into figure of 8 and most of the ganglia concentrate in the anterior part of the body. The pleural and pedal ganglia fuse together and lie immediately beneath the cerebral ganglia forming a ring around buccal mass.
3. The alimentary canal which was straight from mouth to anus, is thrown into a loop.
4. The mantle cavity was primitively posterior in position but after torsion the mantle cavity opens just behind the head and its associated parts are shifted forwards.

How torsion occurs?

Thomson (1958) recognized seven ways in which torsion occurs:

1. Complete or 180⁰ rotation is achieved by muscle contraction alone. It is known only for *Acmaea*.
2. 180⁰ rotation achieved in two stages, first 90⁰ movement by contraction of larval retractor muscle and later 90⁰ rotation by different grwoth e.g. *Patella, Haliotes*.

3. 180^0 rotation by differential grwoth process alone e.g. *Vivapara*
4. Rotation by differetial grwoth process, with anus coming to a position appropriate to the adult state as in *Aplysia*
5. Torsion no longer recognisable as a movement of viscero - pallium e.g. *Adalaria*
6. The visceral mass and shell were dorsal now becomes ventral by torsion.
7. The original symmetry which the animal possessed before torsion now lost. The ctenidium, ospheridium, auricle and kidneys more developed on one side then other. In some specilized froms (*Pila*) the ctenidium, kidneys and auricles etc, are present on right side and disappear from left side but in few primitive forms like *Patella* the symmetry is prevented by the presence of two gills, aurciles and kidneys but the right kidney serves as gonoduct.

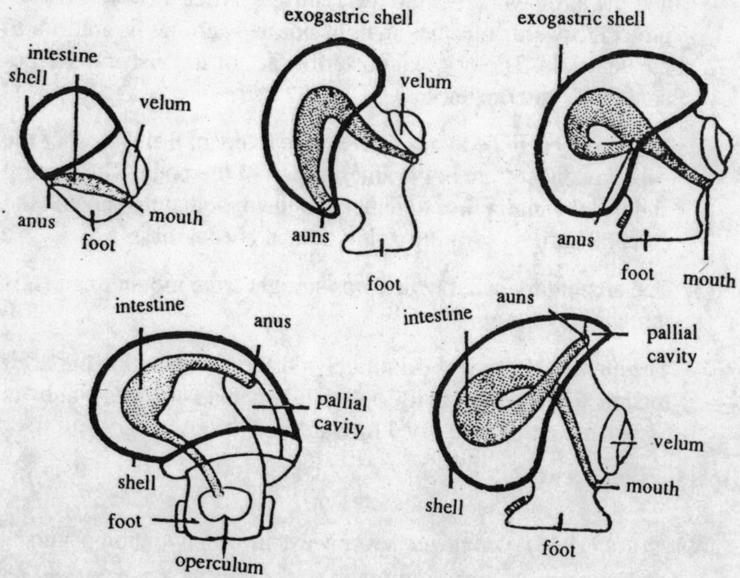

Fig. 10.1. Five successive stages showing development of a gastropod to explain torsion. A. Early embryo without ventral flexture. B. Embryo with an exogastric shell in a lateral view. C. Fig. showing 90° of a lateral anticlock-wise torsion with endogastric shell while mantle and anus moving towards right side. D. 90° torsion stage (Posteriorly). E. Adult form with complete torsion (180°) in lateral view.

Theories of Torsion

There are various theories to explain the casue of torsion a few are described here:

1. Torsion occures in the embryo due to antogonism between the growth of foot and that of shell (*Banton* 1919)

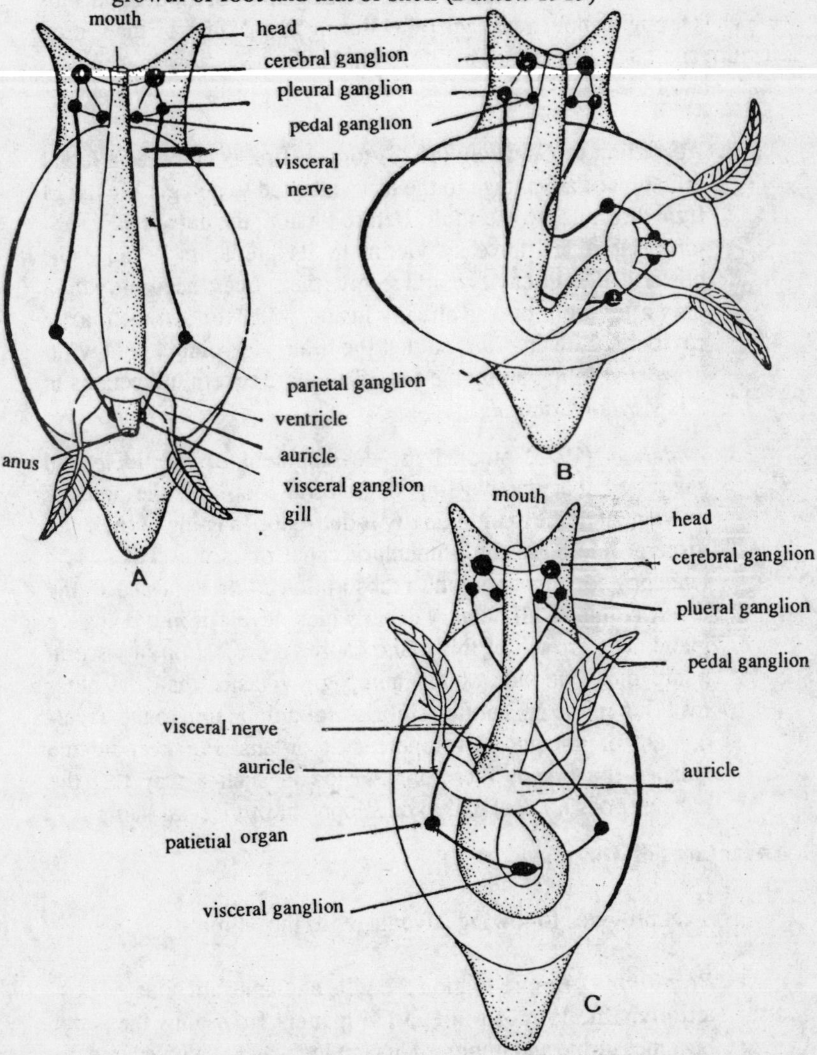

Fig. 10.2 Process of torsion.

Due to development of one piece protective shell, which encloses the soft parts inside it, opens on one side through which the head and gills are to protrude out and for this the head as well as the posterior extrimity of the canal have to be brought about on the same opening of shell. This requirement shifted the ctemidia and mantle cacity to come towards the head end. This is also existed by the extensive development of the foot. The developemnt of foot results in the shifting apart of mouth and anus which is brought about near each other during larval stage by the ventral flexure and lateral flexure. Therefore the foot is necessary for the phenomenon of torsion.

2. According to *Garstang* (1928) torsion first occured as a larval mutation of advantage to the larva adapted to pelagic life but of little direct use to the adult. Before torsion, the untwisted swimming larva fell an easy victim to its predators because the posterior mantle cavity could receive the delicate head and velum only after the foot was already inside. After torsion the mantle cavity became anterior, so that the head and velum could withdraw first followed by the foot. There are several objections to this view of *Garstang*.

3. *Anderson* (1925) studied the development of *Paulidine* and suggested that the twisting of visceral loop and the anterior movement of the mantle Cavity is due to the twisting of shell and visceral hump. In the alimentary canal of mollusc there is a tendency for digestion and reabsorption to be confined to the diverticulum of alimentary canal which develops into digestive gland. The growth of this gland caused a projection in visceral hump thus one side of the hump grow faster than the other resulting in the upset of equilibrium of animal due to the development of unequal development of organs. For keeping the balance the organs therefore twisted in such a way that the balance does not lost due to development of visceral hump.

Advantages of Torsion.

Torsion provides following advantages to the animals:

1. *Respiration.* Before torsion the gills and anus are situated posteriorly. Due to release of faecal matters from anus the water becomes turbid and impure. This impure water is inhaled into the

mantle *cavity* for respiration. After torsion the gills are brough forward. Hence, by torsion, the animal gets the previlage of taking pure water for respiration.

2. *Locomotion.* Before torsion the respiratory current and water current oppose each other. This creates difficulty to locomotion. After torsion, gills are placed forwards and the respiratory water is taken from ahead. So the respiratory current concides with the water current . This does not produce any hindrance to locomotion.

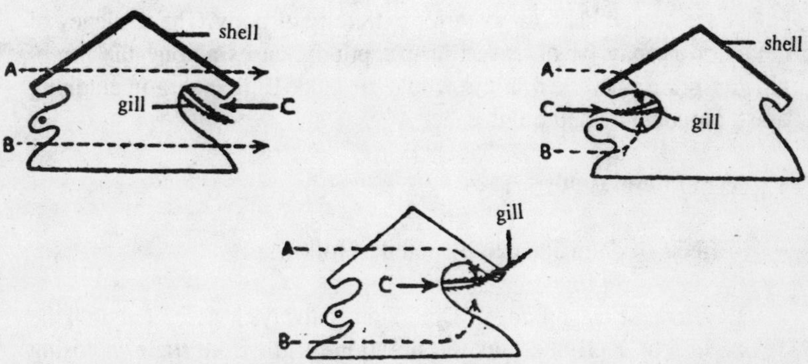

Fig. 10.3. Showing the advantage of torsion.
A. Current due to flow of water. B. Current due to movement. C. Respiratory current.

3. *Protection.* Before torsion the foot is withdrawn into the body first. After foot, the head and ctenidia are taken in. But after torsion the head and ctenidia are taken in first and only lastly the foot is withdrawn. This gives protection from enemies.

Disadvantages.

Due to torsion following disadvantages are seen:

1. *Sanitation.* Torsion brings the anus close to the mouth, thus the faecal matters released from the anus gives rise to the sanitation problems.

2. *Respiration.* The faecal matters is released forward. It makes the water lying ahead impure which is inhaled for respiration.

Detorsion

Changes occuring in torsion are reversible and these changes are known as *detorsion*. The anus and other organs of pallial complex are carried back to its original position, which is posterior, by detorsion. Thus the lost symmetry reappears to forming symmetrical animal.

In least specialized opisthobranchiats and Pulmonata the torsion is not complete and the pallial aparture is carried only to right of lateral side not to posterior end but in most forms like Aplysia the anus and the pallial cavity are moved back to posterior extremity of body. The tendeney of detorsion is may be observed in exceptional cases among the Streptoneura but it is marked in the whole group of Euthyneura in cludidng opisthobranchiate and pulmonate.

Effects of detorsion

Effects of detortion can be noted as follows:

1. *External morphology.* Detorsion is always asssociated with the loss of shell and thus the libration of gill from their enclosing case. Once the shell is lost the gills become exposed to external current and the anterior position of gill is no larges so great an advantage.

The degree of detorsion can be correlated with the extent of which the shell is reduced (Pteropoda and Tectibranchs).

2. *Mantle and Pallial complex.* Detorsion has created condition in the mantle complex of opisthobranchs that are difficult to understand. According to analysis of *Perrier* and *Fischer* (1911) the left side of mantle roof, taking various structures with it, has been showed to the right in such a way as to becomes part of the mantle floor, fusing with the neck of animal. Thus some structures that are borne on the mantle roof in prosobranchs come to lie in the mantle in opisthobranchs. Also in the detorsion process a right posterior part of the mantle becomes detached from the shell and hangs loose as an inferapallial lobe that extands posteriorly and may attain considerable size as in *Philine* and other cephalaspids. In *Akera* the posterior edge of the inferapallial lobe bears a long tentacle as also in *Aglaja*. In some cephalaspids, notably the

Actenoidae, the mantle is drawn out posteriorly into a conspecuous caecum which bears internally a dorsal and a ventral ciliated tract concerned in maintaning a current through the mantle cavity. The caecum may project freely, often spirally rolled, or may be fused to the surface of visceral mass. The mantle of opisthobranchs generally lacks siphons but in *Aplysia* and other anaspids the anus is located on a projecting spout of the posteriolateral corner of mantle called by *Haffmann* (1932-1939) the cloacal siphon, site of an out going current. A true mantle sipon may be substituted by temperory rolling into a tube by the posterior margin of the cephalic shield as in *Ringicula* and *Gastropteron* as already noted.

3. *On alimentary* canal. The alimantary canal before detorsion and after torsion makes coil but after detorsion it again comes to its orginal positon and becomes straight as it was before torosion. This gives the symmetry in animal.

4. *On respiratory* system. Due to detorsion the shell is lost and the respiratory organ such as gill or ctenidia are librated. The gill becomes directly exposed to external current. The anterior position of gill is no longer so great and advantage and that is why the gills are situated in the posterior region on either side of the anus in detorted forms.

5. *On nervous system.* The pulmonate central nervous system is of detorted type with the ganglia concentrated into a ring around the alimentary canal. Usually this circum-enteric ring is located behind pharynx so that the oesophagus and the ducts of the foregut glands must pass through it, but in a few cases it encircles the pharynx anteriorly. The detorsion is brought about by the migration of the intestinal and visceral ganglia forward to contact with the pleural ganglia. Thus eleminating the pleuro-visceral connectives. These connectives are still evident in *Chilina* and traces of them remain in some other primitive pulmonates (*Otina, Ellobium*). Not only detorsion results in the formation of symmetrical nervous system but an advanced concentrate type of nervous system. It may not be a direct effect of detorsion but it follows the sequence of symmetrical development of nervous system.

6. *Effect of detorsion on development.* The new idea put forward by *Thompson* (1958, 1962) about detorsion is against the whole con-

cept of torsion and detorsion in which we believe in the views like of *Pelseneer* (1906-1911).*Pelseneer* stated that in detorted forms the primitive torsion is manifest in the course of development in that in the larvae the mantle cavity is anterior and dorsal. But this was denied by *Thompson* (1958-62) who noted no indications of torsion in the larvae of two doridecean nudibranchs whose development he studied nor in other nudibranch larvae examined. It appears that in opisthobranch development there is generally no recaptilation of torsion, but the organs are formed in their definitive sites. However, a posterior displacement of the anal complex is not uncommon. Thompson himself noted that at metamorphosis of Adalaria the anal complex moves back along the right. Side of its definative posterior position, thus reversing torsion. During the development of the onchidacean *Onchidella* (*Fretter*, 1943) the right side rotates to a dorsal and the mantle cavity with its contained openings (anus & nephridiopore) migrates back along the right side. In the Pleurobranchid, *Berthellina Citrina (Gohar and Abul-Ela* 1957) the right side rotates to a dorsal position, the dorsal side shifts to the left, and the anus is carried back. A posterior displacement the anus also occurs during the development of nudibranch *Eolidina citrina (Tardy*, 1962). These changes may be regarded as detorsional since the main effect of torsion is to bring the mantle cavity and contained complex to an anterior dorsal position. Such a position of the mantle cavity does not seen to occur during the development of opisthobranchs. In nudibranch veligers the mantle cavity is generally located lateroventrally on the right side (*Thompson,* 1962).

11

NERVOUS SYSTEM IN MOLLUSCA

The molluscs prossesses a well developed nervous system which usually consists of a number of ganglia which are linked by commissures and sending out thread-like nerves which ramify into the various organs. The pattern of nervous system varies greatly in different groups of Mollusca. From extreme complex condition to a simplest form in which ganglia are almost entirely absent.

I. Nervous System in Monoplacophora

The nervous system consists of a circum oral nerve ring or cerebral commisssure surrounding the mouth. From it arises two pairs of nerves in posterior region, the inner *pedal cords* and outer *visceral cords*, which join posteriorly to form loop. the pedal and visceral cords are connected by 10 pairs of *transverse connectives*. Nerves extend from the cords to the gills.

II. Nervous System in Amphineura

The nervous system is primitive type. The nervous system consists of cerebral ganglia which are placed above the mouth on either side.

The pedal ganglia are situated on either side below the oesophagus or at the base of foot. The position of pleural ganglia varies considerably but always below the oesophagus and slightly above the pedal ganglia.

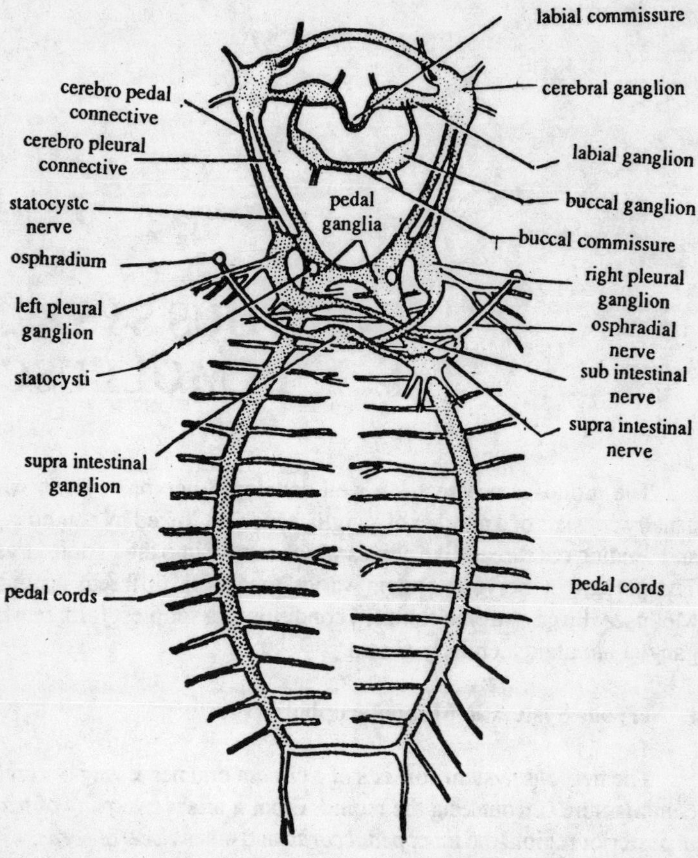

Fig. 11.1 *Patella*. Nervous System.

In chitons there are four nerve cords parallel one another and present nearly whole length of the mass. The exterior parts represents the pallial cords and the inner, the pedal cords. the two pedal cords are united by transverse connectives. The two pallial cords are united posteriosly by a thick supra-renal commissure. Ganglia are lacking or at least poorly developed.

In *Neomenia* there is a distinct cerebral ganglia formed by the massing together of the two nerve cords on each side.

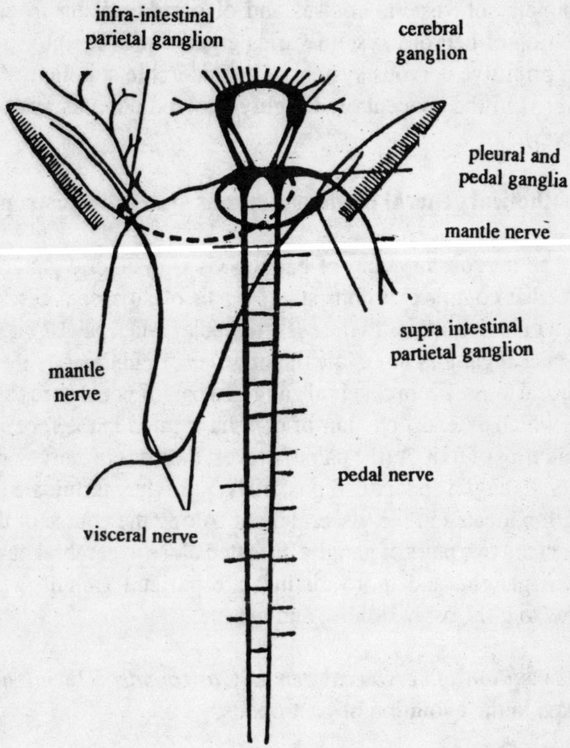

Fig. 11.2. *Halistis*. nervous System.

III. Nervous System in Scyphopoda

The nervous system is symmetrical. A pair of each cerebral and pleural ganglia lie close together on the dorsal side of oesophagus. The pedal ganglia are present in foot and united to the cerebral and pleural ganglia. Lying on either side are visceral ganglia, united by pleuro-visceral connectives to the pleural ganglia.

IV. Nervous System in Gastropoda

With the exception of endoparasitic Endochonchidae all gastropods possess a well developed nervous system in which cerebral, pedal and visceral nerve centres with their respective connectives and commissures are present. The special character of gastropod nervous system is asymmetry of visceral centres and of nerves arising from them. The condition of nervous system varies greatly in different groups ranging from primitive nervous system of considerable simplicity (*Patella* and *Haliotis*) to the concentrated highly evolved nervous system (*Limnea, Tethyes).*

Hypothetical general plane of nervous system in gastropods.

The nervouss system of gastropods is distinctly ganglionated and somewhat complex. It consists of a pair of adjascent cerebral ganglia which give rise nerves to the eyes, tentacles and a pair of buccal ganglia. Thé buccal ganglia innervate the muscles of radula and other structures of buccal mass. From cesebral ganglia a pair of pedal cords arise posteriorly, which extends to a pair of ganglia located in the foot and supplies the muscles of foot. A fine pair of nerves, the visceral nerves, arising from cerebral ganglia and extend posterirly until they terminate in a visceral ganglion located in the visceral mass. Along the course of these nerves are located two pairs of ganglia. Located close to cerebral ganglia are the pleural ganglia and more distinct are parietal ganglia which supply nerves to gills, ospheridium, and mantle.

Evolution of nervous system in Gastropods. Three tendencies are marked in the evolution of gastropods:

1. The entire nervous system has evolved along two definit lines due to torsion, in one type of nervous system the loop formed by the visceral nerve is persisted over itself forming a figure of 8 i.e. streptoneuran system present in prosobranchiates. While in other case the loop remains untwisted i.e. euthyneuran type mostly present in Opisthobranchiates and pulmonates.
2. Secondary the nervous system in gastropods shows a marked tendency towards the concentration of various ganglia and in certain pulmonates all the ganglia form a ring around the oesophagus while finally in *Tethys* the ganglia of one side fuse with each other.
3. Thirdly the long pedal cords persist in the primitive cases of *Patella and Haliotis* are lost during the evolution and finally the pedal ganglia come to lie near the other ganglia without having

long pedal cords.

Types of nervous System on the basis of cerebropleural connectives. There catagories have been recognized:

1. *Hypoathroid.* In this case the pleural and pedal ganglia are quite distinct and the pleural ganglia lie in intimate contact of the anterior part of pedal ganglia and the result is that the cerebro-pleural and cerebropedal connectives are quite long and nearly equal to one another *e.g. Patella.*

2. *Dysthroid.* The pleural ganglia, in this case, move toward the cerebral ganglia, with the result that cerebro-pedal connectives are longer in comparision to those of cerebropleural connective *e.g. Ampularia.*

3. *Epiathroid.* In this case pleural ganglia fuse with cerebral ganglia and the result is that cerebro-pedal cocnnectives are quite long while cerebro-pleural connectives are lost. *e.g.* various pectenibranchiates.

Nervous system in subclass Streptoneura. The visceral commissures are in the form of 8.

Nervous system in order Aspidobranchia. Here the concentration of ganglia is little. In most primitive streptoneura the cerebral ganglia are placed at the sides of oesophagus apart from each other. They are connected by a cerebral commissue. The pedal ganglia give out too long pedal cords extending along the ventral side united to transverse connective. The pleural ganglia are approximated to the pedal ganglia. Somtimes an infra or labial commissure is also present. *e.g. Patella, Haliotis* etc.

Nervous system in order Pectinibranchiala. The nervous system is somewhat concentrated. Without a labial commissure except in *Paludina* and *Ampullaria.* The nerve collar is situated behind the buccal mass except in *Ampullaria.* The number of visceral ganglia varies from 1-3. The pedal cords and transverse commissures are absent except in *Paludina and Cyclophorus*.Progressive zygoneury (supra or sub-intestinal ganglion from which all the mantle nerves of that side appear to be given out) is present in majority of cases in which it is present. It takes place on right side. In other pectinibranchiates sometimes dialoneury (mentle is supplied by nerves of pleural ganglion of same side as well as by a nerve from supra-intestinal ganglian of same side) is present. In *Cassidaria* zygoneury as well as dialoneury is present. In *Natica* where the anterior part of foot is strongly developed, a pre-pedal ganglion becomes differentiated from the pedal ganglion.

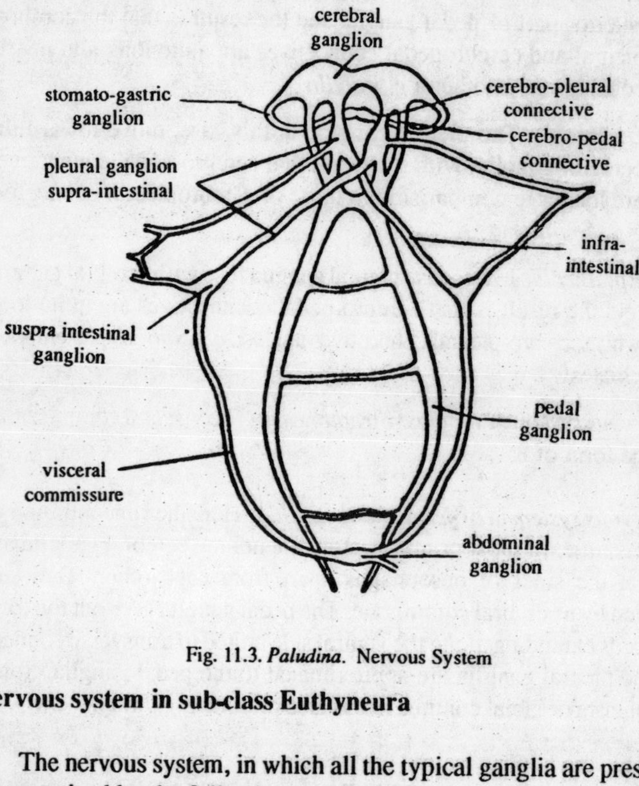

Fig. 11.3. *Paludina*. Nervous System

Nervous system in sub-class Euthyneura

The nervous system, in which all the typical ganglia are present, is characterized by the following peculiarities:

1. By the absence of streptoneury except in *Acteon*.
2. By the marked tendency towards the concentration along the posterior end of pharynx.

The animals of this group are symmetrical forms having untwisted pleuro-visceral connectives due to detorsion. The external opening and mantle cavity are shifted from anterior to the posterior side and thus regain symmetry secondarily. The ganglia tend to become in a ring like form united by short connectives and commissures. *e.g. Aplysia.*

As a type we can see the nervous system of *Bulla* the following points:

1. Pleural ganglia have shifted till they lie close to cerebral ganglia resulting in the shortening of cerebro-pleural connectives.
2. Three visceral ganglia.
3. Parapodia are innervated by pedal ganglia.

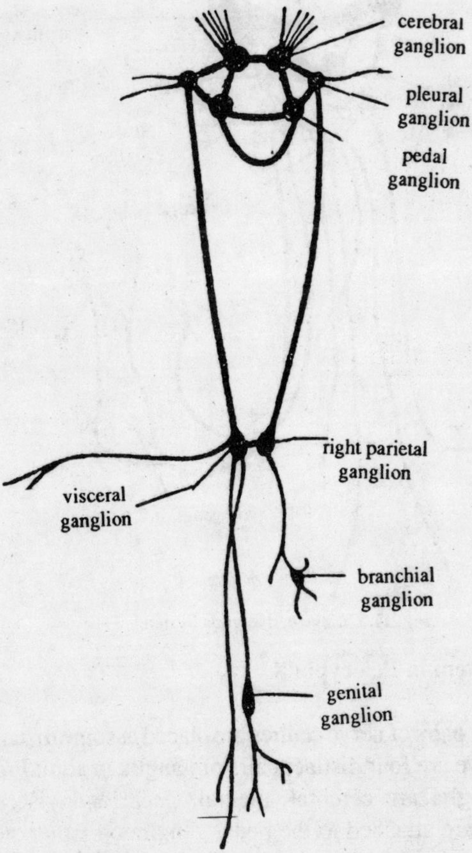

Fig. 11.4. *Aplysia.* Nervous System

Nervous System in sub class Pulmonata.

All typical ganglia are present here. They are put together immediately behind the pharynx and is present a peri-oesopheageal complex into which even the parietal and visceral ganglia have been drawn. The cerebral ganglia lie close on dorsal side, while all other ganglia lie ventrally and the cerebro-pedal and cerebro-pleural connectives are easily distinguishable. The buccal ganglia are always found lying posterior to the pharynx below the oesophagus. e.g. *Limnea*.

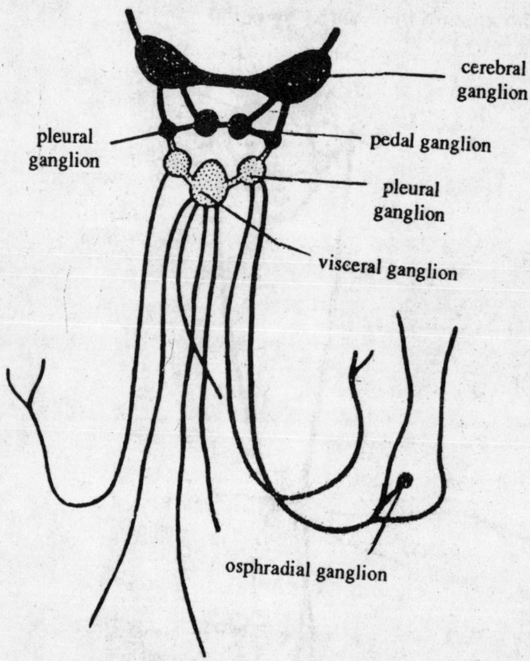

Fig. 11.5. *Limnea*, Nervous System.

V. Nervous System in Pelecypoda

The different pairs of nerve centres are placed at some distance from one another. There are four distinct pairs of ganglia in adult Protobranchus. These ganglia are cerebral, pleural, pedal and visceral. The cerebral ganglia are attached to the pedal ganglia on either sides thus having two pedal connectives on each side i.e. cerebro-pedal and pleuro-pedal as in *Solenomya and Nucula*.

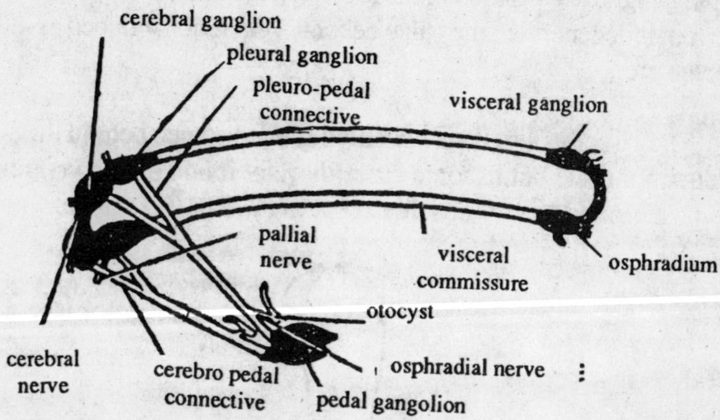

Fig. 11.6. *Nucula*. Nervous System

In all the pelecypods the cerebro-pleural centres are connected with the pedal and visceral centres, therefore, two types of connectives are present the pedal and visceral connectives. In *Mytilus* the visceral connective is fused for a certain distance with pedal connectives.

In *Protobranchia and Venus* the cerebro pleural ganglia are attached to one another but in all other groups they are apart from each other. The

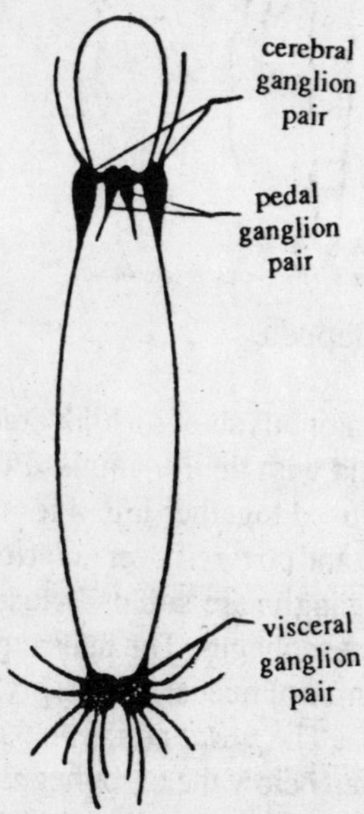

Fig. 11.7. *Pecten*. Nervous System.

pedal ganglia are situated in the foot been attached to one another. When the foot is degenerated they also become very much reduced as in *Teredo*, *Pecten* etc.

The visceral ganglia are situated some distance behind the posterior adductor muscle but in some cases they are found on the ventral side of the posterior adductor muscle except in *Pholas*, *Teredo* etc.

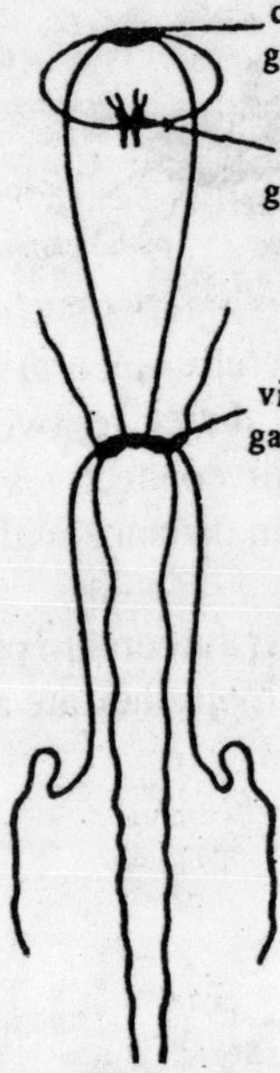

Fig. 11.8. *Terdo*. Nervous System.

VI. Nervous System in Cephalopoda.

The nervous system of cephalopods shows a high grade of development. It consists of paired ganglia with their commissures and connectives. The cerebral ganglia are fused together into a rounded mass and protected by the cranial cartilage and covered over anteriorly by a strong fibrous membrane. The buccal ganglia are situated close to the buccal mass on the anterior side of the oesophagus. The buccal ganglia may be differentiated into superior and inferior buccal masses giving the appearence of separate cerebral ganglia. The *pedal ganglion* is in the form of an united *ganglionic mass* situated below the oesophagus and is divided into an anterior lobe or branchial ganglion giving nerves to the arms and a posterior lobe or *infundibular ganglion* supplying the funnel. From the *branchial ganglion,* ten *branchial nerves* to the arms are given off, therefore, these arms are regarded as a part of the foot. *Branchial ganglion* is also connected to the superior buccal ganglia by paired *branchio-buccal commissures* and to the cerebral ganglia by *cerebro-branchial commissures.*

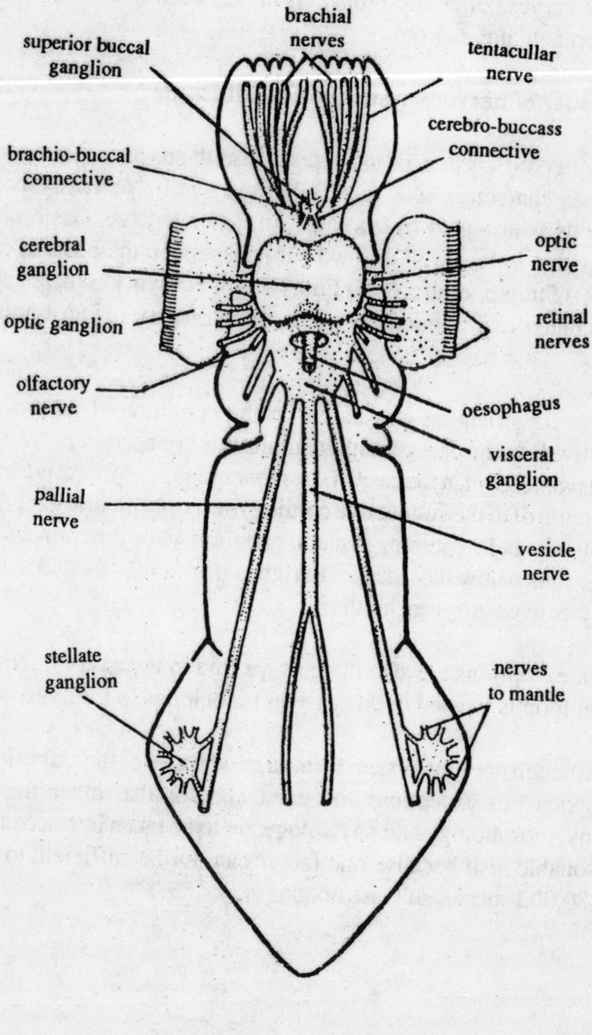

Fig. 11.9. *Loligo.* Nervous System.

A pair of *Pleuro-visceral ganglia* in the fused state are found situated posterior to the *infundibular ganglia*. From this ganglia are given off two stout *pallil nerves* to the mantle, where each dilates to form a large *stellate ganglion*. From this *stellate ganglion* nerves are given off to the mantle. From the *pleuro-visceral ganglia* originate paired visceral nerves which supply internal organs. From the visceral nerves originate a pair of nerves called the *branchial nerves* each of which expands to form a branchial ganglion.

Importance of nervous system in Classification.

The nervous system of molluscs is also the basis for its classication like other characters like foot, gills and shell. No classification of gastropods is accepted by zoologist in general. Class Gastropoda has been divided in two well defined sub-classess on the basis of nervous system (i) Streptoneura and (ii) Euthyneura. Euthyneura includes opisthobranchiates and pulmonates while Streptoneura includes prosobranchiates.

This classification was based on the condition of pleuro-visceral connectives. In the case of euthyneurans this connection is symmetrical and not twisted, but in the case of streptoneurans the connective is twisted in the figure of 8. the connective coming from right pleural ganglion runs over intestire before joining visceral ganglion while the connective from left side runs below the intestire to right side of the body and in this way the connectives cross each other.

But exceptions are also there as we find in the case of *Acteon*. The visceral loop is twisted in this case but it belongs to Euthyneura.

Although nervous system forms a good basis for the classification of Gastropoda but exceptions too exist and for that other factors like anatomy, morphology and physiology are to be taken into account and it is reasonable also because one factor can not be sufficient to classify about 30,000 species of Gastropoda.

12

PEARL AND PEARL INDUSTRY

A pearl is a conscretion produced by a mollusc. Its culture is paying induustry these days. The pldest record of using pearls go back to the chinese people. They are said to have used them as for back as 2300 B.C. Royal families and nobles of every country in age have adorned their persons with pearls.

What is a pearl ?

A pearl is a concretion formed by mollusc. It consists of *nacre* or mother of pearl as the mollu's shell. It is secreted by the mantle, the inner muscular lining of the shell, as a means of protection against foreign body. A faultless pearl in formed by the deposition of nacreous substances arrabged one after the other in a concentric layer and laid down by the epithelium of mantle.

Pearl formation

The pearl formation is an adaptation against outside materials. When a foreign material such as a sand-grain or a prarsite happens to enter the body it adheres with the mantle. The mantle epithelium immediately grows over the material in the form of a sac and enclose it. This mantle epithelium start secreting concentric layers of nacre around the foreign material. The complete structure is called pearl.

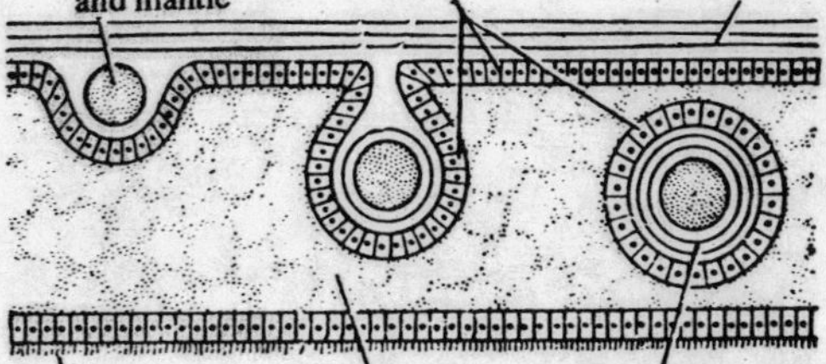

Fig. 12.1. Pearl formation

Composition of Pearl

The pearl is formed by nacre. The nacre is formed of two substances, the calcium carbonate which is in the form of *argonite or calcite* and and albuminoid substance called *conchiolin* ($C_{30} H_{42} N_2 O_{11}$). The chemical analysis of a typical pearl is:

Calcium Carbonate	88-90%
Organic matter	3.8 - 5.9%
Water	2-3%
Residue	0.1-0.8%

Types of Pearls

Following are the four main types of pearls:

1. *Blister pearls.* When some foreign matter such as and, worms, crabs, fish or pieces of shell becomes lodged between the fleshy mantle and the shell blister pearls are formed, such pearls are attached to shell.
2. *Baroque pearls.* There pearls are most likely to be perfectly spherical, for they are formed inside the flesh of mantle where succeeding layers of nacre may be added on all sides.
3. *Hem pearls.* These are formed near the margin of mantle where

the dark pigments usually produce brown or blackish pearls.

4. *Seed pearls.* The small pearls are called seed pearls.
5. *Cultured pearls.* These are the pearls obtained from cultured species of pearls oyesters.

Pearl producing animals.

Pearls are produced by bivalve molluscs. These are marine as well as fresh water:

1. The fresh water mollusc is:

Unio margaritifera.

2. The marine molluscs are:

(a) *Pinctada Vulgaris*
(b) *P. fucta*
(c) *P. Chemnitzi*
(d) *P. margartifera*
(e) *P. anomioides*
(f) *P. atropurpurea*
(g) *Haliotis*
(h) *Mytilus*
(i) *Placuna blacenta*
(j) *P. maxima.*

The species of mollusc which is mainly is *Pinctada vulgaris.*

Culture of pearls

The culture of pearls is a complex but sensitive process. It is involves the following steps:

1. Collection of Oysters

(i) *Japanese Method of Oyster Culture.* The larvae of the oyster freely swim in the shallow waters of the sea they are called *spats.* They are maintained in nurseries upto a certain time. For collection of spats different methods are employed. They are

(a) Semi-cylindrical tiles.

(b) Bamboo poles with branches

(c) Shells of various molluscs

(d) Plastic nests

(ii) The oysters are collected from the bottom of the sea.

(iii) In the laboratories the eggs of pearl oysters are fertilized and young ones are obtained.

2. Preparation of graft tissue

The piece of tissue which is inserted into the oyster is called graft-tissue. The graft must be in the form of a square of 2 x 2 mm in size and usually it is cut off form the mantle of another oyster.

3. Preparation of nucleus

The nucleus is a foreign material which is inserted into the oyster and is in the form of a bead 2 mm in diameter.

4. Implantation

The oyster's foot is exposed and a small incision is made on the foot. On this incision the graft tissue in placed. The nucleus is placed on the graft tissue. Then the oyster is released in the cage.

5. Rearing of oysters.

The operated oysters are placed in cages and are suspended in the sea from rafts. This type of *roft culture*. Rafts are rectangular wooden frames floating at the water surface. They are made to float by empty diesel drums. The rafts is kept in place by anchors. Besides raft other methods are there, these are long-line method, rack culture, pole culture and bottom-culture method.

6. Harvesting.

Pearls attain their maximum size in three years and after the oysters are removed from cages and the pearl is taken out.

13

LARVAL FORMS IN MOLLUSCA

In molluscs the development may be direct or indirect. In the direct there being no larval stage where the young ones hatching from the eggs resemble their parents in general appearence, except the size. The direct development is seen in cephalopods. In indirect development a larval stage is seen during the development in majority of mollusc classes.

Paludina is a viviparous form. In some cases like *Lottorina, Cenia* etc. the metamorphosis takes place inside the egg case. Howeover, three main larval forms are met with in Mollusca. These are:

1. Trochophore
2. Veliger
3. Glochindium.

1. Trochophore.

The ciliated trochophore in its typical state appears in *Patella and Trochus* and in these it is formed at a very early period, even before the formation of the mesoderrm, and becomes free at once. It is pear shaped and measures about 0.5 mm in length. In some cases they are free where as they are enclosed in egg. A circlet of preoral cilia, the *prototroch or velum* divides the body into two unequal parts, the upper one consist of prostomium where as the lower part bearing mouth and anus. The pre-oual part is large and convex with an *apical plate* bearing the long cilia called the *apical cilia* as seen in the larvae of *Patella, Dentalium* etc. Near the apical cilia there are two ciliated elevations each consisting of a single cell. At the posterior end there are large anal cells bearing a bunch of cilia called *telotroch*. Between the ciliary girdle and anus, on the dorsal side, the area of ectoderm constitute the *shell glands* which secretes a thin, saucer-shaped larval shell. The trochophores of Lamellidae consists of an additional shell inside the first one which is cast off later. A slight ridge in the neighbourhood of shell represents the border of mantle. Alimentary canal is made up of single layered epithelial cells in the form of a tube. It comprises mouth, stomodaeum, oesophagus, stomach and intestine (mesenteron). A single large cell gives rise to the mesoderm. A rudiment of redula is present in oesophegus.

The trochophore contains two cerebral ganglia, a pair of eyes from the underside of apical plate.

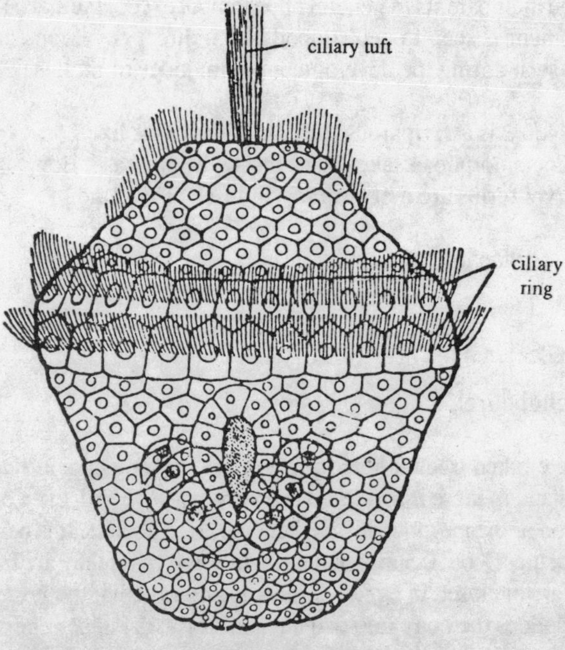

Fig. 13.1 Trochophore of *Patella*.

The statolith sacs appear as depression of ectoderm at the sides of the mouth. Soon the trochophore develops into a yet more efficient locomotor form, the *veliger*.

2. Veliger larva.

The preoral ciliate area or *velum* begin to protrude on both sides as a bilobed flap. The velum are very delicate and extensive, and a very delicate organs of locomotion. Between the bases of velar lobes the anterior end of the larva is provided with eyes and tentacles. The larva has a shell into which the velar apparatus can be withdrawn and which encloses the partly developed viscera. Alimentary canal is complete and the anus is shifted to anterior side. A foot usually bearing an operculum in present. In addition to velar apparatus, there are some organs like larval heart and kidney also present. The larval kidneys are paired and generally symmetrical organs situated at the anterior end of the body immediately behind the velum. Statocyst and gill-rudiments are also present.

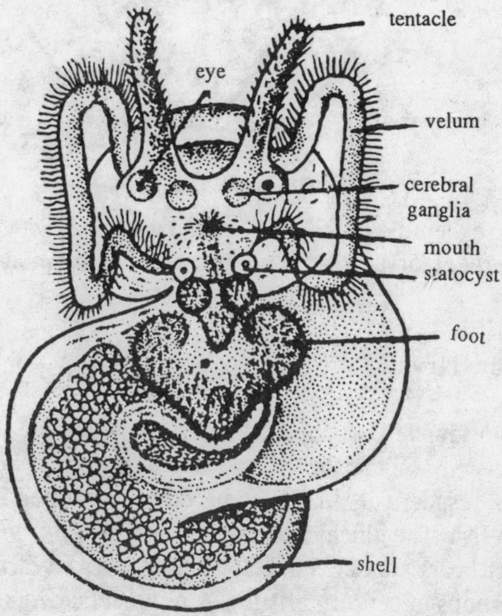

Fig. 13.2. Veliger larva of *Vermetus*.

Some gastropods have feeding veligers (planktotrophic) with a larval life that may last as long as three months. Others have a brief yolk-laden nonfeeding veliger (lacithotrophic). The long cilia of the velum function not only in locomotion but also in suspension feeding. During the course of the veliger stage, torsion occur and the shell and visceral mas & twist 180° in relation to the head and foot.

In most of the pulmonates, the young are hatched out with the characters of the adult. In the species in which veliger stage is passed in the egg, the velar lobes are reduced and the larva rotates in the egg with the help of velar cilia.

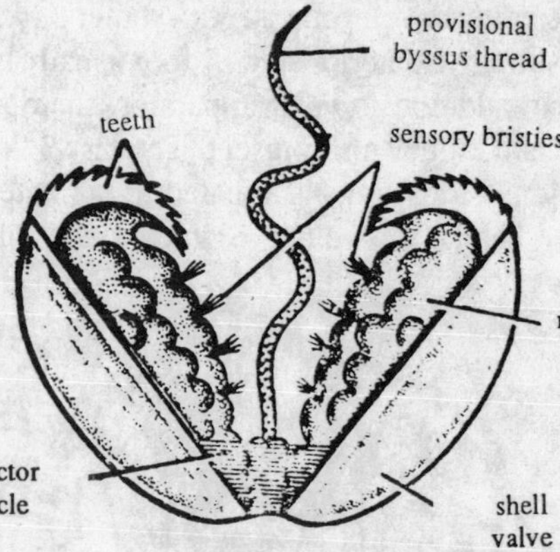

Fig. 13.3 Glochidium larva of *Unio*

With the exception of *Dreissena and Nausiforia*, which have a free-swimming veliger larva, fresh water bivalves exhibit modifies development.

3. Glochidium larva.

In the worldwide Unionacea, the African Mutilidae, and the South American Mycetopodidae the veliger, called as *glochidium, haustorium and lasidium* respectively, has become highly modified for a parasitic existence on fish. The glochidium larva is enclosed by two valves, each edge of which bears a hook. The shell valves cover a larval mantle, which bears four groups of sensory bristles. A rudimentary foot is present, to which is attached a long adhesive thread, the *byssal thread.* There is neither mouth nor anus.

When mature it measure from 0.1mm to 0.5mm depending upon the species. In *Unio and Anodonta*, the glochidium leave the gills through the suprabranchial cavity and exhalant aperture. In *Lampsilis*, the glochidia emerge directly from gills through temporary openings. The glochidia of *Anodonta* immediately clamp into the fins and other parts of body surface of the fish. The glochidia require certain species of fish as host; others can tolerate a wide range of host species and the byssal thread aids in initial contact and adhesion (*Wood*, 1974). The larval mantle contains phagocytic cells that feed on the tissue of host and obtain nutrition for the developing clam. The parasitic period lasts from 10-30 days. In the mean time the parasite is surrounded by the over growth of skin of fish forming a cyst.

Some of the larger freshwater mussels may produce as many as 3,000,000 glochidia.

14

ORIGIN OF MOLLUSCA

The position of Phylum Mollusca is in several respects are considerable and positive isolation and any attempt to derive it from one group or other in the view of major evidences to lay any hypothesis for the origin of Mollusca is ill founded. It should be noted with interest that even in such remote time as Cambrian Period, the sponges, echinoderms, Platyhelminths and mollusc have already identify themselves as a separate and distinct group with the same distinction from each other what we observe in the present time. In such circumstances any system of classification or any theory of development which proposes the derivation of mollusca from any of this group is extremely danagerous.

It would, therefore, almost certain that the origin of Mollusca is merely a metaspeculation and if at all any attempt is to be made we have to examine the indication of the relationship of the phylum by the consideration of embryonic resemblance. One of the most important feature of phylum Mollusca is the occurence of a *trochophore larva* in the development. It is interesting to note that such a larval form is also shared by annelids but if the presence of trochophore is in any form to be regarded as important in shading some light from the origin of Mollusea, it should not necessarily be taken as indicative of an annelid ancestor. If we examine the trochophore larva of any mollusc we see a complete gastrulation and the ciliated bands besides a single large cell, which would give rise to the mesoderm in the successive developmental stages, is seen. In addition to the position of an apical organ i.e. *protroch* and larval nephridia its internal organisation also shows similarities with the annelidian trochophore. Such apperent similarities appeal to be discussed and a more careful study would be certain striking differences between the two which can not escape our attention. We thus say on the dorsal side of advanced molluscan trochophore there is a rudiment of mantle formed by ectoderm. At this stage the earliest sign of the shell can also be seen.

The foot has also made its appearence as a ventral prominence. The single large mesoderm cell gives rise to two bands of mesodermal cells in each of which a cavity develop, but by for the most important difference from annelid is afforded by the absence of segmentation in the embryonic mesoderm of mollusc. It would thus appeal that absence of segmentation provides a positive evidence from such group of animals which were devoid of segmentation.

From the foregoing account it becomes positively certain that we have to leave our such view for the origin of Mollusca from some annelid type ancestor and possibility of deriving the group from some turbellarian ancestor may help us. Such a connection is proved by the striking similarties that are present between the ladder-like nervous system in certain turbellarians and in placophore, solenogasters and certain aspidobranch gastropods. In either case we find that the ganglia itself-are distributed along the entire length of the main nerve trunk. The pleuro-visceral and pedal nerve cords should be taken to correspond to the lateral and ventral nerve cords of the turbellarians. Now if such a hypothetical ancestor could suppose to has given rise a primitive mollusc then we should have to postulate following important changes that should have undergone in its organisation for such a purpose:

1. Secretion of dorsal shell which in the begining might have been in the form of a layer of cuticle containing calcareous particles and would have served as a protective covering.
2. The development of shell would evidenty reduced the general respiratory surface of the body which in turn would make the necessity to develop such specialized respiratory organ as ctenidium as present in molluscs.
3. The ventral musculature which have already been well developed in the turbellarian must have undergone a remarkable development to give rise to such a specialized and efficient molluscan foot.

It is worthy of notice that typical trochophore larva bears close resemblances to certain form of turbellarian larva i.e. *Muller' s larva.* It we have to assume that the primitive molluscs derived from some turbellarian like ancestor we will also have to postulate the conversion of Muller's larva into a molluscan trochophore. Such a postulation would autometically lead us to the contentation that the annelids and mollusc had probably common platyhelminth origin will subsequent extreme divergence into two phyla. If ontogeny repeats phylogeny then a trochophore will be taken to be present a remote ancestor of annelids and molluscs which resembles the ancestors of platyhelminthes represented by Muller's larva. The trochophore larva by certain changes developed into aneliden form. These changes includes:

1. Elongation of body which becomes segmented.
2. Formation of septa.
3. Development of coelom.

On the other hand transformation into molluscan form occured through the following changes:

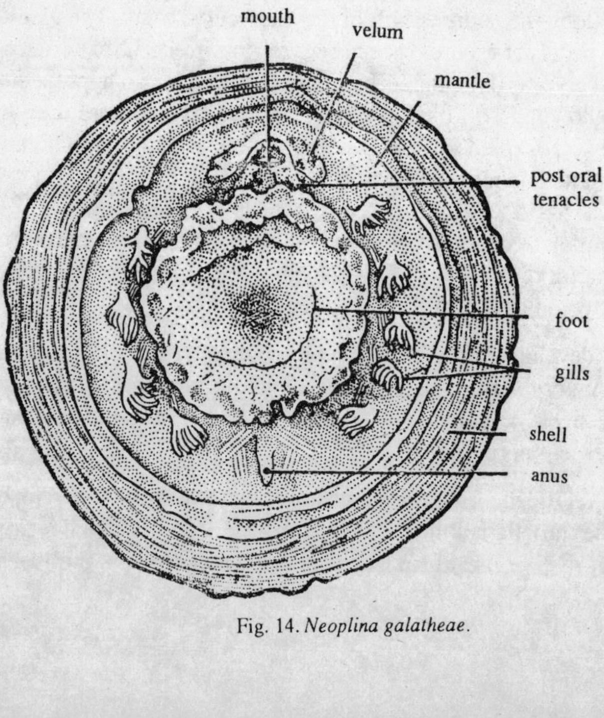

Fig. 14. *Neoplina galatheae.*

1. The apical cilia got transformed into velum which serves not only for locomotion but also for feeding. The cilia of apical plate create a current of water which bring food particles into the mouth.

2. Shell gland got developed which secretes the unique shell.

But in 1952 the whole picture changed when 10 living. specimens of *Neoplina* were found from a depth of 3500 metres off the Paciffic coast of Costa Rica (Maxico). The most unusual feature of *Neoplina* which belongs to the Monoplacophora of Mollusca is its appearence, metameric plane of structure. It displaces 5 pairs of gills, 6 pairs of nephridia and these organs are more or less metamerically arranged. a number of zoologists believe that this metameric plane displaced from *Neoplina* is an evidence that Mollusa comprise a fundamental metameric phylum and must have evolved from the annelids. The molluscan trochophore is thus belived to be additional evidence of annelid origin of mollusc.

In the light of *Neoplina* it may existed to revise our thinking in future, but at present the older thing assuming a common origin of annelid and mollusc seems to possess the greatest weight of evidence.

15

EVOLUTION OF VARIOUS CLASSES OF MOLLUSCA

From the primitive archimollusc structure different groups, known today, have diverged and each of the groups, follows the evolution along certain lines of specialization.

1. Amphineura or Placophora.

The Amphineuran are the Chitons, seem to have departed least from the ancestral mollusc. Their body becomes elongated and was protected by eight shell pieces on the dorsal side. On either side of mantle cavity was arranged symmetrically several pairs of bipectinate gills. If we were to assume that the primitive mollusc possesses several ctenidia then it is appearent that the chitons have retained the primitive character. On the other hand if we assume that the primitive mollusc posses a single pair of ctenidia then it naturally employes that the numerous ctenidia of chiton are formed by the repetation of the former. The foot in chiton has retained the primitive from. The nervous system is also primitive in that the ganghonic cells are distributed along the whole length of nerve cords. Chitons also possess a redula which is well developed.

2. Gastropoda

The leading features in the evolution of Gastropoda has been the torsion of visceral mass producing symmetry in body. There is no doubt that the ancester gastropod possessed a body with a straight alimentary canal, leading in a posteriorly placed anus and on either side of anus was placed the ctenidium. We also know well that gastropods have univalved shell which is covering all the soft parts of body and extension of foot on the ventral side tend to remove the pallial complex with and renal openings away from head. The univalved shell, therefore, ment apperently necessary for mantle cavity, ctenidia and excretory aperture to shift at anterior and somewhat in the neighbourhood of opening of shell. The whole of pallial complex was shifted from its primitive posterior position to the anterior position by the torsion. The large majority of gastropods included in order Prosobranchiata shows torsion in full development. It is also known that each group of Mollusca shows a progressive loss of shell in the course of evolution. It has also been invariably observed in Gastropoda that such a loss of shell is accomplished by a reversion of the phenomenon of torsion i.e. detorsion and consequent rotation of visceral loop back to its posterior condition i.e. in opisthobranachiates. The Opisthobrnachia are derived from pectinibranchia is prooved by the fact that they possess a single ctenidium, single auricle, besides embryological evidences also support such a view. Even those opisthobranchs which are detorted in adult condition the primitive torsion is seen in the course of development. The nudibranches seem to have more step further. Here the detorsion is complete and have lost the ctenidium as well as the shell. But they have evolved from a fully torted gastropods when we study their embryology.

We have tried to show a gradual evolution of different groups of gastropoda on the basis of torsion and its subsequent reverser but nervous system also shows a way to the evolutionary history of Gastropoda. We find the most primitive type of nervous system in this class in exhibited by certain aspidobranchs is that they show a lesser tendency towards the concentration of various ganglia, besides they possess longitudinal pedal cords which can compare to pedal cords of Chiton but the evolution of nervous system in this class has taken along two lines:

1. In the establishment of progressive euthyneury are symmetry.
2. In the concentration of various ganglia and their connectives around the circum-oesophageal nerve ring. The visceral loop remains twisted and in all the prosobranchiats are the result of torsion.

The opisthobranchiates shows detorsion have an untwisted visceral loop, a concentration of various ganglia thus begin in the cephalaspides for example in *Bulla* the pleural ganglia shift very close to the cerebral ganglia or they may be fused as in *Aceton.* In *Notochus* the pleural-visceral connectives are considerable shortened. In the nudibranchs the tendency of concentration of various ganglia and shortening of various connectives have gone still further. Taking for example in the case of *Tethys* we find that the pleural, pedal and cerebral ganglia of each side fuse together in the form of compect pleuro- cerebral, pedal mass and with its counter part on other side it forms a large supra-oesophageal ganglionic mass.

So far we have excludes the pulmonates for any discussion. The typical member of Pulmonata shows a coiled visceral mass possess a shell and exhibit an internal streptoneural asymmetry to certain extent but this should not lead us to believe that these are a group of less specialized form. On the other hand they exhibit a perfect symmetry of nervous system, the concentration of various ganglia and shortening of various connectives. The final result is, therefore incorporation of visceral loop together with its ganglia and circum-oesophageal nerve cord.

On an examination of respiratory organs one finds a gradual loss of ctenidia as we move from lower groups to the higher ones in class Gastropoda. We thus see that certain aspidobranchs (*Haliotis*) exhibit the most primitive condition as they retain a pair of bipectinate ctenidia. Other aspidobranchs belong to family Trochidae, Tuberidae and Neritidae have only one ctenidium but that is bipectinate. From this bipectinate type of gills arose monopectinate type of gills of pectinibranchi which have the last leaf-lets of one side of the animals. Certain land prosobranchiates have lost their ctenidia in relation to terrestiral life and consequently respire with the help of pulmonary sac. In *Pila* while still retained the monopectinate gill has developed a pulmonary cavity and is capable of leading an amphibious life. When a true ctenidium is present among opissthobranchiates it lies posteriorly on right side as a result of detorsion. The nudibranchs have lost the true ctenidium as well as mantle cavity and respiration is carried with the help of secondary gills or outer skin. Finally in pulmonata the ctenidia are totally absent and the respiration is effected with the help of pulmonary sac.

So for as the foot in gastropods are concerned, in aspidobranchia it is very well developed and is differentrated into propodium, metapodium and epipodium. All of which show remarkable development. In pectinibranchia epipodium is usually absent, in certain prosobranchiates the foot may be reduced in simple discoidal prominence (*Vermetus*). It may be represented by small, ventral appendage in parasitic forms. Passing to Opisthobranchia we find that there is no division of foot of various forms. The parapodia may be modified into fold-like extensions from the edges of the sole of foot and bent back over the shell. Among certain nudibranchs the foot no longer exists as a differentiated organ while pulmonate show poor development.

From what has been said above it can be safily concluded that the most primitive gastropod, aspidobranchia, which in tern gave rise to the pectinibranchia. The pectinibranchia by a series of changes in their internal as well as external organisation gave rise on one end to the Opisthobranchia and on the other Pulmonata and have made specialization among two different lines.

3. *Pelecypoda*

As far as bivalvia is concerned the fact that bivalved shell at first appearence is univalve and that of the foot of Protobranchiata is of creeping type and the ctenidia are plume or feather-like suggest the derivation of their class from animal resembling a simple gastropod with no redula and undisturbed bilateral symmetry. On the other hand certain zoologists record this proof to have been derived from the hypothetical primitive ancestor with a more or less developed head and redula. Such a primitive mollusc with the process of degeneeration on account of leading an inactive sedentary life was modified accordingly either lost or fail to acquire those specialized characters of the organization of highly developed forms. The pelecypods thus arose with absence of any definite region and masticatory apparatus.

From the point of view of physiology, the most primitive pelecypods are those in which the foot has a creeping ventral surface and the branchial filaments are free and not reflected those of the protobranchiate. From these are evolved the filibranchiates where branchial filaments are reflected but are still lacking vasscular function. These is their term gave-rise to the eulamellibranchiates which are most specilized in respect of comp lication of Ctenidia, Numerous vascular interlammellar and inter filamentar junctions are present. From eulamellibranchiates are derived septibranchiates in which respiratory organ is represented by branchial sephum.

The most primitive condition of nervous system exhibited by Protobranchiata as the cerebral and pleural ganglia are still separated and there is a set of cerebro-pleural, cerebro-pedal and pleuro-pedal connectives on each side. In all other lamellibranchiates, however, the cerebro-pleural ganglia are intimately fuse to one another and there is a single pedal connectives on each side. The two pedal ganglia also lie close to each other and may be reduced in those forms in which the foot is reduced or absent *e.g. Pecten, Teredo, Ostrea* etc.

4. *Cephalopoda.*

The cephalopods are bilaterally Symmetrical and have a posterior mantle cavity. The primitive form in this group i.e. tetrabranchiates possess an external shell which may be many chambered besides they possess two pairs of ctenidia, kidneys and auricles in contrast to more specialized form i.e. Dibranchiata which have each of these structure in single pair and have an internal shell which may be reduced or absent. The remarkable features of cephalopods, however is extra-ordinary modification of foot into arms and tentacles. This class remains apart from mollusks by its wonderfully high organisation, specialized nervous system and eyes. There is nothing to indicate close relationship with any of lower groups beyond the general resemblance to molluscan organisation andpresence of odontophore.

The cephalopods form, infact, a singular ray isolated group. Palaeontology has also not given any indication of their origin. Embyrology is equally silient. The absence of free-larva and performed modification in the development produced by enormous mass of food-yolk sharply separates them from all the members of Phylum Mollusca.

Index